CHEMISTRY RESEARCH AND APPLICATIONS

NEW DEVELOPMENTS IN LIQUID CRYSTALS RESEARCH

CHEMISTRY RESEARCH AND APPLICATIONS

CHEMISTRY RESEARCH AND APPLICATIONS

NEW DEVELOPMENTS IN LIQUID CRYSTALS RESEARCH

ANKA TRAJKOVSKA PETKOSKA
AND
ANITA TRAJKOVSKA-BROACH
EDITORS

nova publishers
New York

Library of Congress Cataloging-in-Publication Data

New developments in liquid crystals research / Anka Trajkovska Petkoska (associate professor, University St. Kliment Ohridski-Bitola, Faculty of Technology and Technical Sciences-Veles, R. Macedonia), Anita Trajkovska-Broach, editors.
pages cm. -- (Chemistry research and applications)
Includes index.
ISBN 978-1-63482-456-9 (hardcover)
1. Liquid crystals. I. Trajkovska Petkoska, Anka, 1970- editor. II. Title: Liquid crystals research.
TA418.9.L54N494 2015
530.4'29--dc23
2015008475

Published by Nova Science Publishers, Inc. † New York

CONTENTS

PREFACE

Liquid crystalline materials were discovered more than 100 years ago, but their profound impact was in the past four decades through successful commercialization of liquid crystal materials into devices that we use in our everyday lives. The research on liquid crystals (LCs) continues actively in 21st century with new proposed LC materials and advanced applications, usually associated with displays, personal gadgets, sensors, optics, lasers, biomedical, and other device concepts.

This book offers up-to-date topics related to LCs intended for a broad spectrum of professional profiles in both, academia and industry, from beginners to advanced scientists, physicists, chemists, materials scientists, bioscientists, engineers, and so on. The scope of the book is to present some of the recent innovations in the field of LC research and their potential in various display and non-display applications.

The book contains six chapters, as follows:

The first two chapters are related to polymer cholesteric LCs (PCLCs) made in form of microparticles of tailored shapes and dimensions. Particularly, *Chapter 1* describes the manufacturing process of microparticles made of PCLC material by use of soft lithography technique. Initially, a silicon wafer as a master is used for making an elastomeric flexible replica of the master that serves as a mold to manufacture the shaped forms of PCLC particles. Variety of instrumental techniques are used for their characterization. *Chapter 2* presents the electro-optical behaviour of regularly tailored PCLC particles when they are placed in a suitable host fluid systems under an applied electric field. Modification of the electro-optical behaviour of PCLC particles by doping with particulate materials of different nature, e.g., conductive and non-conductive dopants, with nano- and micron-structures, is presented here. Possible applications of PCLC microparticles are proposed in this chapter as well.

Chapter 3 follows the research on microparticles (also, called flakes) and describes their microencapsulation and manipulation under an applied electric field. COMSOL MULTIPHYSICS software in combination with an analytical model is used to investigate several microencapsulation configurations in an applied AC electric field. The modeling predicts that test cells with multiple materials in the electric field path can have a wide range of electro optic responses in AC electric fields; PCLC flakes' reorientation occurs due to Maxwell-Wagner polarization.

Chapter 4 presents a novel study of combination of nanomaterials and liquid crystals. The effect of carbon nanotubes, quantum dots, and other nanoparticles, as well as their derivatives on the physical properties of LC nanocomposites, is investigated. The properties, such as the threshold voltage, dielectric anisotropy and elastic constants are known to be of key importance for liquid crystal display applications.

Chapter 5 gives an overview of structural and magnetic properties of four different types of metallo-mesogens, particularly iron-containing liquid crystals. These materials are of special interest, because they are combining the properties of LCs and metals. Electron paramagnetic resonance (EPR) and Mössbauer spectroscopy were used for characterization of the magnetic behavior of these LC compositions.

Chapter 6 provides an overview of alignment technologies for nematic liquid crystals, proposed and developed in the past two decades. Rubbed polymer and obliquely-deposited inorganic films, as well as photo-irradiated, IB-bombarded and patterned surfaces, are briefly reviewed along with several novel alignment concepts. Reports related to the molecular origins of a preferred LC orientation on these surfaces are highlighted with an attempt to address the long-lasting debate on the relative contribution between surface-mesogen interactions and surface topography.

It is my belief that this book will be a great asset to the large pool of existing literature related to liquid crystalline materials and their applications. It is envisioned that this book, as a collection of chapters proposing novel LC materials and LC applications, will inspire a wide range of researchers and engineers towards new explorations of the unique properties of liquid crystalline mater and its behavior under externally applied fields.

February 2015

Editors:

Anka Trajkovska Petkoska, PhD
University St. Kliment Ohridski-Bitola,
Faculty of Technology and Technical Sciences,
Veles, R. Macedonia
Email: trajkovska_anka@yahoo.com;
anka.trajkovska@uklo.edu.mk

Anita Trajkovska-Broach, PhD
Scientific and Technical Consultant
CSI: Create. Solve. Innovate. LLC
Christiansburg, VA, USA

In: New Developments in Liquid Crystals Research
Editors: A. T. Petkoska and A. T. Broach
ISBN: 978-1-63482-456-9

Chapter 1

TAILORED POLYMERIC CHOLESTERIC LIQUID CRYSTALLINE MICROPARTICLES

Anka Trajkovska Petkoska*

University St. Kliment Ohridski-Bitola,
Faculty of Technology and Technical Sciences-Veles, R. Macedonia

ABSTRACT

Liquid crystalline phase is a special condensed matter phase that belongs between (disordered) isotropic liquid and (ordered) solid crystalline phase. Liquid crystals (LCs) possess the fluidity of a liquid phase, and at the same time, exhibit an orientational, and a certain degree, positional order. In general, LC materials consist of rod-like or disk-like molecules. The anisotropic molecular shape and the presence of a certain degree of order are responsible for the significant anisotropic optical, electrical, magnetic, and mechanical properties, which find many unique applications in optics, electronics, and photonics. According to the microscopic order and the interaction between molecules, there are different classes of LC phases: *nematic, cholesteric and smectic.*

Cholesteric liquid crystals (CLC) are materials that exhibit unique optical properties, such as selective reflection and circular polarization due to their helical structure. With proper molecular design, polymeric CLCs with varying pitch lengths that reflect color from deep UV to far IR region can be made.

Polymeric CLC (PCLC) materials have been used in forms of *films and particles.* Due to their polymer nature, PCLCs have the ability to be processed easily into large-area, monodomain films, which can find potential applications as notch filters, reflectors, polarizers, and a variety of novel displays. *PCLC irregular particles* have been proposed for many passive and active optical applications, as well. The challenge is making PCLC microparticles of tailored shapes and sizes, which will result in a well-controlled response of the microparticles under an applied external field. Therefore, the topic of this chapter is the manufacture of *tailored PCLC particles.* A soft lithographic technique has been used for the manufacture of regular PCLC microparticles and has been reported earlier by the author. In particular, an inexpensive, flexible elastomeric mold, such as PDMS mold has been used for making a variety of micron-sized PCLC shapes, viz. rectangular,

* Corresponding author address; Email: anka.trajkovska@uklo.edu.mk.

square, oval, etc. Shaped PCLC microparticles, made by a soft lithographic technique, have been characterized by different techniques, viz. polarized optical microscopy, scanning electron microscopy, white light interferometry and other techniques.

Among many potential applications, the most attractive application of regularly-shaped PCLC microparticles is as the low-cost alternative to the existing liquid crystal displays (LCDs), and particularly, the particle display technologies, due to their multi-color availability and low power consumption.

Keywords: Polymer cholesteric liquid crystals, regularly shaped PCLC microparticles, soft lithography, characterization of PCLC microparticles

1. Introduction

Liquid crystals (LCs) represent a special condensed matter phase, which belongs between the disordered isotropic liquids and the well-ordered solid crystals. They possess the fluidity of a liquid phase, and at the same time have a certain degree of the orientational and/or positional order like in a crystal lattice. LC materials show pronounced anisotropic properties due to the rod-like or disk-like nature of their molecules. Due to all these properties, LCs posses high anisotropic optical, electrical, magnetic, and mechanical properties [1-3].

There are different types of LCs phases: *nematic, cholesteric and smectic* depending on the microscopic order and the interaction between molecules. A unit vector **n** or *director* describes the preferred orientation of the long molecular axes. *Nematic* LCs possess preferred orientational (long range) order of the molecules (described with **n**), but no positional order. When the director of the molecules lies perpendicular to the plane containing the molecules and the molecules keep their orientational order within the layer, the phase is called *smectic* A. Hexagonal (positional) ordering of the molecules within the layer gives smectic B phase. If the director makes an angle other than 90° to the plane, then smectic C phase is formed. In the smectic phase there is molecular order within the layers, but not between layers. *Cholesteric* (or *chiral nematic*) *LCs* have the orientational order of a nematic LC within the layer and the local director rotates in a 360° helical fashion around an axis perpendicular to the layers with a period known as *pitch length* (*p*). The rotational direction defines the handedness of the cholesteric phase—left or right-handed [1,2,4,5].

The *selective reflection*, one of the most important optical properties of cholesteric LCs (CLCs), is determined by the pitch length, the helical twist sense and the refractive index of the cholesteric material. "Bragg-like" selective reflection from a CLC film consisting of many layers occurs at certain wavelengths that satisfy the relationship: [6,7]

$$\lambda_0 = \bar{n}p[\cos\frac{1}{2}\{\sin^{-1}(\frac{1}{\bar{n}}\sin\varphi_i) + \sin^{-1}(\frac{1}{\bar{n}}\sin\varphi_s)\}] \quad (1)$$

$$\lambda_0 = \bar{n}p\cos\varphi \text{ for } \varphi_i = \varphi_s = 0 \quad (2)$$

where:

λ_o is the wavelength at which the reflection occurs [8],

$\bar{n}$ is the average refractive index given as:

$$\bar{n} = \sqrt{(n_e^2 + 2n_o^2)/3} \quad (3)$$

where:

n_e and n_o are the extraordinary and ordinary index of refraction of the nematic sublayers, respectively,

p is the pitch length of CLC material,

φ_i and φ_s are the angles of incidence and observation, respectively.

The bandwidth of selective reflection, $\Delta\lambda$ is determined by the birefringence of the material, Δn, expressed as:

$$\Delta\lambda = p\Delta n \quad (4)$$

where the material birefringence is given as:

$$\Delta n = n_e - n_o \quad (5)$$

When circularly polarized light having the same handedness as that of cholesteric film is incident on the cholesteric stack, it will be totally reflected and will maintain the handedness of the incident beam, while circularly polarized light of the opposite sense will be transmitted. Outside the selective reflection band, light incident on a cholesteric film will be unaffected regardless of its polarization state.

The color of a CLC film depends on the chemical structures of the liquid crystalline molecules, the orientations of the molecules, the nature and concentration of any dopants, the thickness of the layer and other factors such as temperature, magnetic and electric field [5].

Polymer cholesteric liquid crystals (*PCLCs*) possess a fixed selective reflection color. The long-chain macromolecular nature of PCLC materials may result in glass-formation, processing into films, and good mechanical and thermal stability.

The helical molecular order can be fixed when the degree of freedom of molecular motion is limited (losing the fluidity). If the molecular order is frozen in the solid state by cooling the material below its glass transition, then *glassy CLCs* (*GCLCs*) is obtained. They can exhibit glass transition temperatures of $\geq 50\ ^\circ C$ [5,9-12].

The unique optical characteristics of CLCs are the selective reflection and circular polarization, so they have potential in many optical, photonic and electronic applications.

Proper molecular design of liquid crystal molecules to give CLC materials with varying pitch lengths from deep UV to IR region, as well as their ability to be processed easily into large-area, monodomain, defect-free films, opens many opportunities for many novel applications, such as polarizers, notch filters, reflectors, display applications, etc. [5,9,13-15].

2. Polymer Cholesteric Liquid Crystal Microparticles

2.1. PCLC Materials

The materials explained in this chapter belong to the group of *thermotropic* LCs or materials where the LC phase appears in a certain temperature interval. (The other class of LC materials is *lyotropic LCs*, when the LC phase is a function of solvent concentration) [1,3].

PCLC material is used for making films or tiny particles (flakes). Non-crosslinkable PCLC material (particularly, *non-crosslinkable polysiloxane CLC*, that is cyclic siloxanes substituted with mesogenic groups that are connected to the backbone by aliphatic spacers) is used in this work. They possess a low degree of polymerization (low molecular weight), which results in a relatively low viscosity. They usually do not crystallize, but instead remain in a glassy state (GCLC) and have a glass transition temperature (T_g) of ~ 50 °C and a clearing point (T_c) of ~ 200 °C [10,11,16,17].

The preparation of continuous PCLC films and freeze-fractured particles (flakes) from non-crosslinkable PCLCs was already explained earlier by many authors. Generally, with a thermal treatment of PCLCs (over their T_g), PCLCs melt and lower their viscosity, which make it possible to align them into a *film* from a dry or dissolved state (e.g., dissolved in toluene or dichloromethane). Films are freeze-fractured into *tiny particles* with liquid nitrogen [16,18-20]. The optical properties of used PCLC material is shown in Figure 1. Namely, Figure 1 presents the optical properties for thin (~ 4 μm thick) films prepared of non-crosslinkable PCLCs, blue, green and red (between uniaxially buffed polyimide coated fused silica substrates sandwiched in a parallel arrangement). Figure 1a shows the transmittance/ reflectance measured with unpolarized light while Figure 1b shows measurements made with circularly polarized light on a UV-VIS-NIR spectrometer. One can see that these materials are *left handed* since LCP (left handed, circularly polarized) light is totally reflected while RCP (right handed, circularly polarized) light is almost totally transmitted (Fig. 1b). Other properties of these PCLCs, such viscosity, solubility, optical properties, etc., are summarized in Table 1 [17].

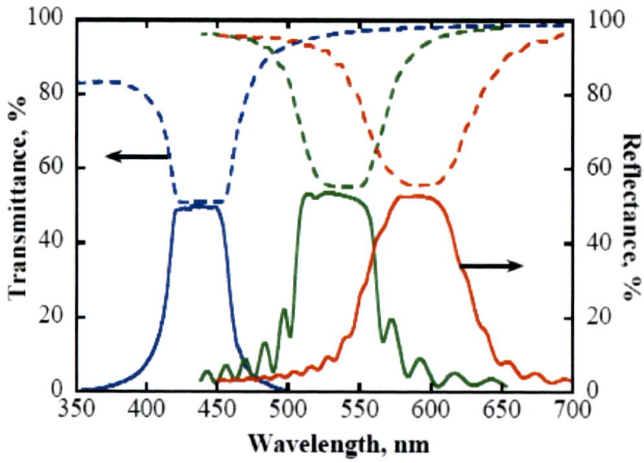

a)

Figure 1. (Continued).

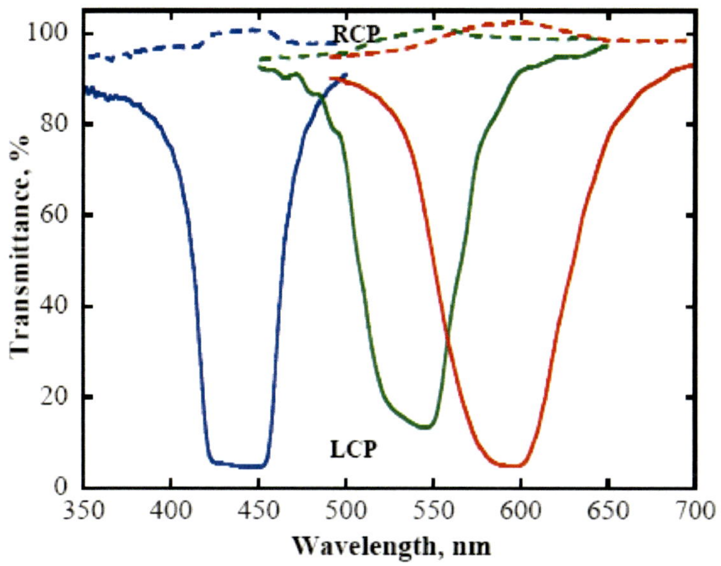

b)

Figure 1. a) Transmittance and reflectance of unpolarized light, b) transmittance of polarized light through PCLCs used in this work (Table 1). (RCP = right handed, circularly polarized light, LCP = left handed, circularly polarized light) [17].

Table 1. Properties of PCLC material

Non-crosslinkable PCLC material	
Manufacturer	Wacker Chemie GmbH (e.g., C4754)
Supplied as	dry, chunks or powder
Composition	polysiloxane
Density, ρ [kg/m^3]	1100
Index of refraction, n_r (at 589.6 nm)†	n_o = 1.5975 n_e = 1.6685
Δn	0.145
Selective reflection wavelength [nm]‡	540
Handedness	left
Bandwidth [nm]	~ 50
Dielectric constant, ε_p	2.2
Conductivity, σ_p [S/m]	1.10^{-9}
T_g [C]	45-55
T_c [C]	180-210
Soluble in	Toluene, methylene chloride
Viscosity, η_p [mPa.s]	0.5-1.5 (at 150 °C)

† index data for nematic substructure, or nematic phase at room temperature.

‡ PCLCs with different selective reflection peaks (blue and red) from the same manufacturer are also available.

2.2. Irregularly Shaped PCLC Microparticles

The manufacturing process for PCLC microparticles (flakes) from a thin continuous PCLC film consists of:

a) knife-coating of melted PCLC material onto a silicon wafer at an elevated temperature,
b) quenching of the PCLC film to the room temperature, and
c) freeze-fracture of the film by use of liquid nitrogen. The film fractures into small microplatelets with random (irregular) shapes and sizes.

These PCLC microparticles can be manufactured to reflect from the UV to the near IR region. Figure 2 shows PCLC microparticles that reflect red, green and blue light. The reflected color, like in CLC films, depends on the refractive index, pitch length of PCLC material and the angle of the observation (shown in equation 1). They possess the same properties as PCLC films, like stability, circular polarization and selective wavelength reflection [13,16,18].

PCLC microparticles are characterized by a variety of techniques: polarizing optical microscopy (POM) (Fig. 2), scanning electron microscopy (SEM) (Fig. 3), white light interferometry (Fig. 4). Also, by use of atomic force microscopy (AFM), pitch length is measure of fractured PCLC microparticles and it was shown that all key structural and optical properties are preserved after flaking [16-18,20].

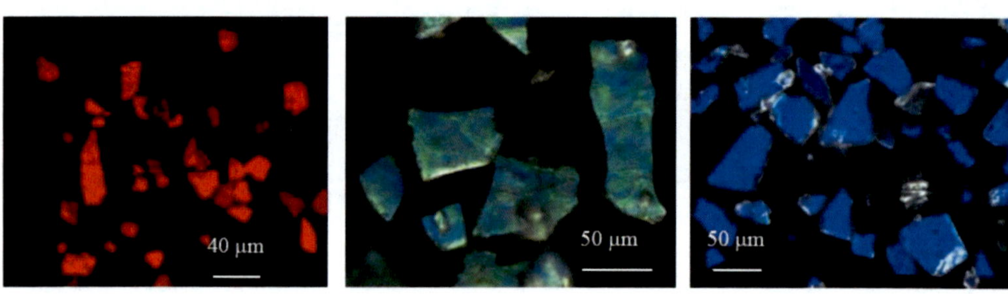

Figure 2. Irregularly shaped PCLC microparticles characterized by polarizing optical microscopy in reflective mode with crossed polarizers (true color).

Figure 3. Irregular shaped PCLC microparticles characterized by SEM.

Figure 3 shows variability in size, shape and thickness for irregular shaped PCLC particles characterized by SEM. The measured particle thicknesses varied from ~ 4 μm to ~10 μm [17].

Figure 4 compares the variability in thickness between two irregular shaped PCLC microparticles (e.g., a particle with ~10 μm and particle that is 15 μm thick), characterized by white light interferometer [17].

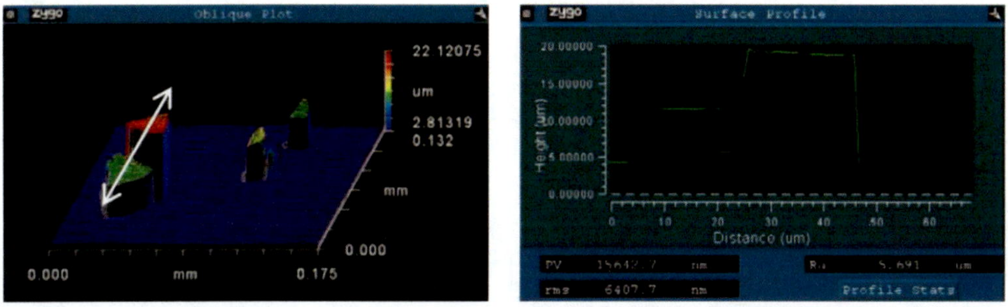

Figure 4. Irregular shaped PCLC microparticles with different thickness characterized by white light interferometry.

2.3. Regularly Shaped PCLC Microparticles

Due to certain disadvantages of irregularly-shaped PCLC microparticles, like random shapes, no control over their dimensions, great variations in thicknesses, sharp edges, it was of crucial importance to shape the PCLC materials/films into controllable tailored microparticles. Particles with known dimensions, shapes and thickness, are easier to control and their electro-optic behavior (active application) is easier to model and predict than that for irregularly shaped microparticles.

After many attempts to make shaped PCLC microparticles, the novel method—soft lithography technique—proves to be an effective method that can make large quantities of PCLC microparticles. In particular, shaped microparticles are made in a flexible polydimethylsiloxane (PDMS) mold. A patterned silicon wafer is used as a rigid template over which an elastomeric material PDMS is cast. Air curing of PDMS over the template usually takes ~24 h at room temperature. After hardening, the PDMS replica is peeled off from the silicon wafer and an inverse replica of the wafer is obtained. This replica is actually a flexible mold for making shaped PCLC microparticles [17,21-26].

2.3.1. Patterned Silicon Wafer

Depending on the design on the patterned silicon wafer, different shapes of PCLC microparticles can be manufactured. Figures 5–7 show different types of characterization of patterned silicon wafers used as templates for making a flexible mold for tailored PCLC microparticles.

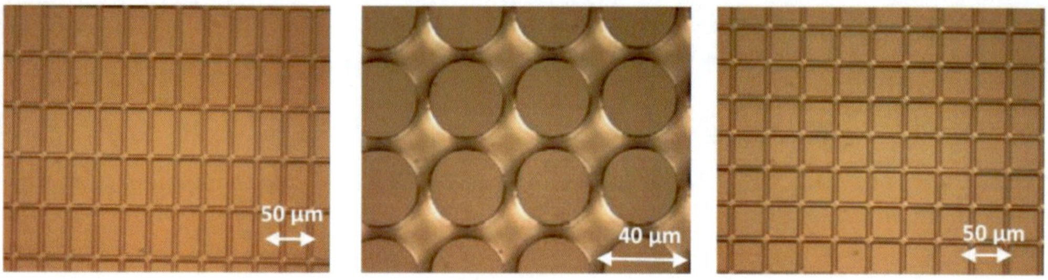

a) b) c)

Figure 5. Silicon wafer with different patterns characterized by POM: a) 60 µm x 30 µm (20x magnification), b) 40 µm diameter (50x magnification), c) 40 µm x 40 µm (50x magnification).

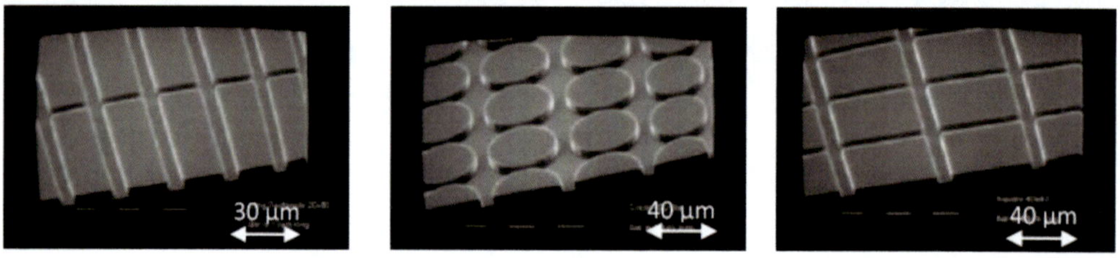

a) b) c)

Figure 6. Silicon wafers with different patterns characterized by SEM: a) 80 µm x 20 µm, b) 40 µm diameter, c) 40 µm x 40 µm.

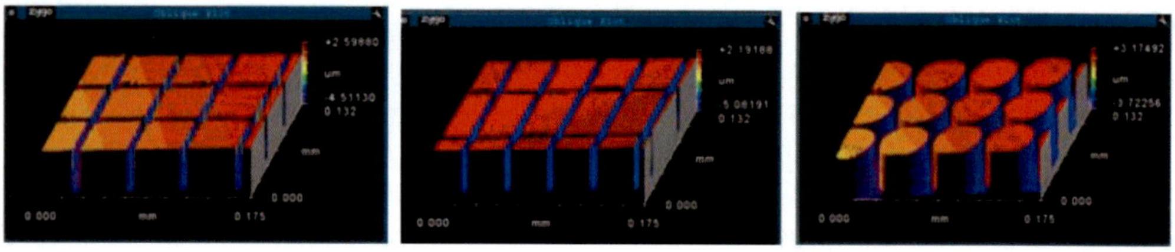

a) b) c)

Figure 7. Silicon wafer with different patterns characterized by white light profilometer: a) 60 µm x 30 µm, b) 40 µm diameter, c) 40 µm x 40 µm.

2.3.2. Flexible PDMS Mold

The process of manufacture of flexible PDMS replica of the patterned silicon wafer that can be used as a mold for tailored PCLC microparticles is shown in Figure 8. (As an example: 60 µm x 30 µm ellipsoidal shaped particles can be made in the mold shown in this figure) [17, 21, 25,26].

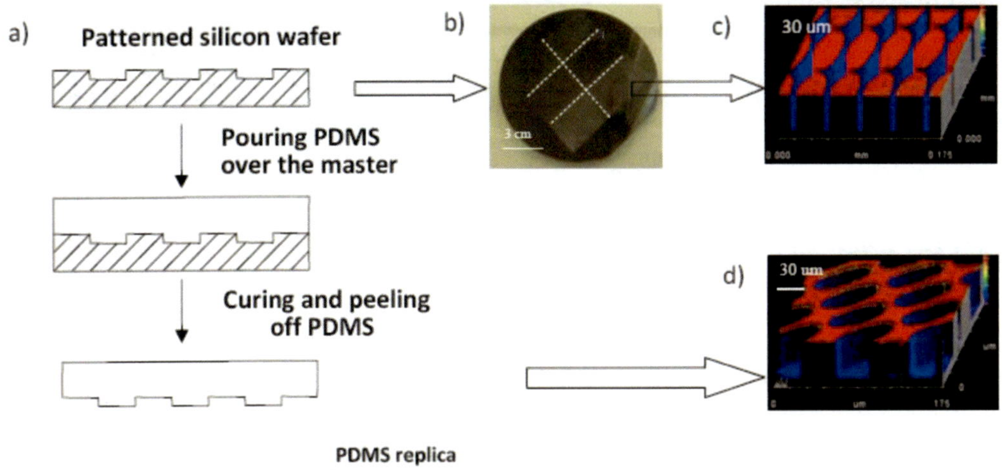

Figure 8. a) Scheme for the process of making a PDMS mold over a rigid patterned silicon wafer, b) patterned silicon wafer (Section 2.3.1), c) ellipsoidal shaped pattern (one of six) on the silicon wafer scanned on white light profilometer, d) white light profilometer image of a PDMS mold as an inverse replica of the patterned silicon wafer shown in c) [17].

The flexible patterned replicas (different PDMS molds) were characterized by different techniques, as shown in Figures 9 and 10 [17,21].

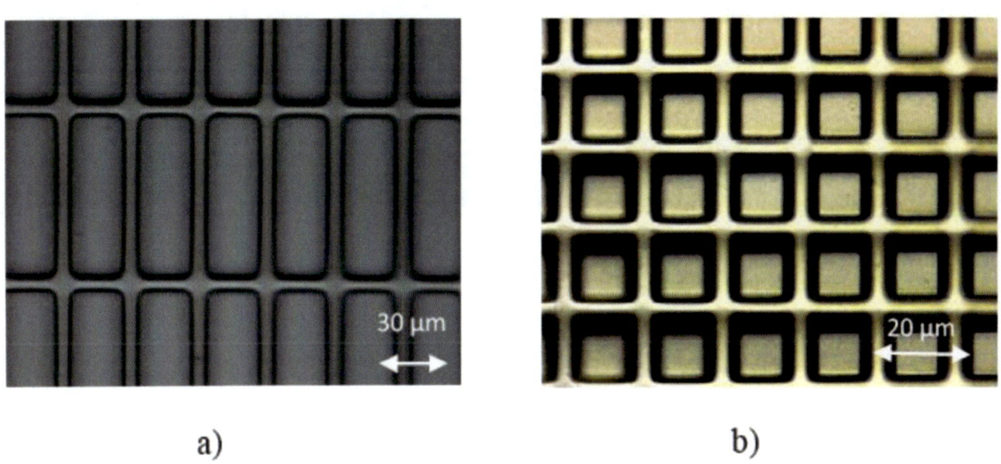

Figure 9. PDMS replicas from patterned silicon wafers characterized by POM (50x magnification): a) 60 μm x 20 μm, b) 13 μm x 13 μm.

2.3.3. Manufacturing of Tailored PCLC Microparticles

The method for manufacturing tailored PCLC particles from non-crosslinkable polysiloxane CLC material is shown in Figure 11 (as an example, a mold replica containing 13 μm x 13 μm x 5 μm square wells is used) [17,21,25,26].

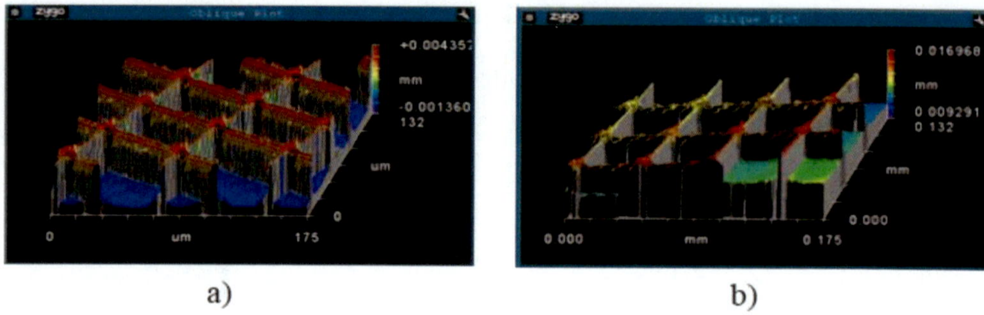

Figure 10. Some PDMS replicas from patterned silicon wafers characterized by white light interferometer: a) 60 μm x 30 μm, b) 40 μm x 40 μm.

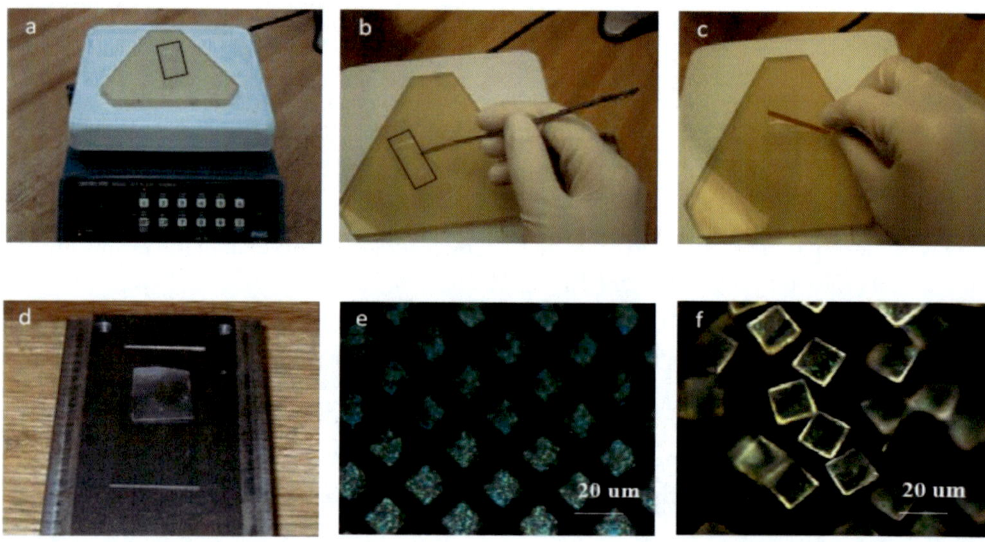

Figure 11. Process for formation of regularly shaped PCLC microparticles: a) preheating the PDMS mold on a hot plate, b) filling the mold with a PCLC material, c) knife-coating, d) quenching, e) examining the level of fill in a microscope, f) bending the PDMS mold and collecting dry PCLC microparticles [17].

The method consists of:

a) placing the PDMS mold on a microscope slide, placing the slide on a glass-ceramic hot plate, and heating the assembly to 130–140 °C on a hot plate;

b) filling the mold with a solution of ~25 wt.% PCLC dissolved in toluene or methylene chloride, or could be a dry PCLC material as well;

c) evaporating the solvent from the filled mold (for manufacturing from the PCLC solution), and "knife coating" the sample with another microscope slide used as the "knife" to establish a good planar alignment;

d) quenching the microscope slide and mold on a metal substrate to room temperature;

e) examining the wells under a microscope to determine the level of fill, and

f) bending the mold over a collecting dish, and ejecting the PCLC particles from the mold.

Steps a) through d) could be repeated in order to either fabricate layered particles or to increase their thickness. The same mold may be cleaned with methylene chloride solvent and reused multiple times to make additional quantities of particles.

3. Characterization of PCLC Microparticles

3.1. Polarized Optical Microscope Characterization of PCLC Microparticles

Under illumination at near-normal incidence, the three colors (red, green, blue) are apparent in the true color images of PCLC microparticles taken with a polarized optical microscope (POM) in reflection mode (Fig. 12).

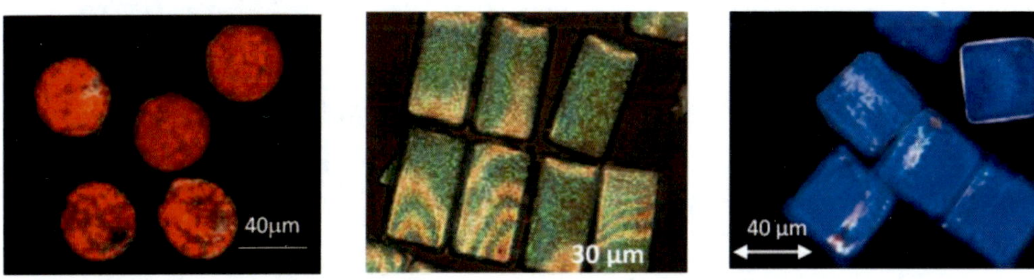

Figure 12. Tailored PCLC microparticles characterized by POM in reflective mode, crossed polarizers.

3.2. Characterization of PCLC Microparticles by White Light Interferometry

Image analysis using white light interferometry (Fig. 13) confirms the lateral dimensions and thickness uniformity of the microparticles (Fig. 13a and b) [17].

a) b) c)

Figure 13. a) Tailored PCLC microparticles characterized by white light interferometer, b) line scan over 4 microparticles designated by the arrow superimposed in a) to show thickness uniformity, c) rectangular shaped PCLC particles show the uniform lateral dimensions from particle to particle [17].

3.3. SEM Characterization of PCLC Microparticles

SEM characterization of tailored PCLC microparticles confirms good uniformity in dimensions and thickness (Fig. 14a). Figure 14b shows SEM pictures of a square particle. The picture on the left side reveals sloped edges on a particle that follows the side-wall contour of the PDMS mold. The picture on the right is a top view of a square particle. In Figure 14c, SEM pictures show a layered structure of microparticles made with multiple fillings (~3 layers, in the picture on the left) and the bottom sides of two particles joined by excess PCLC material (these particle surfaces were in contact with the mold surface, right side of Fig. 14c).

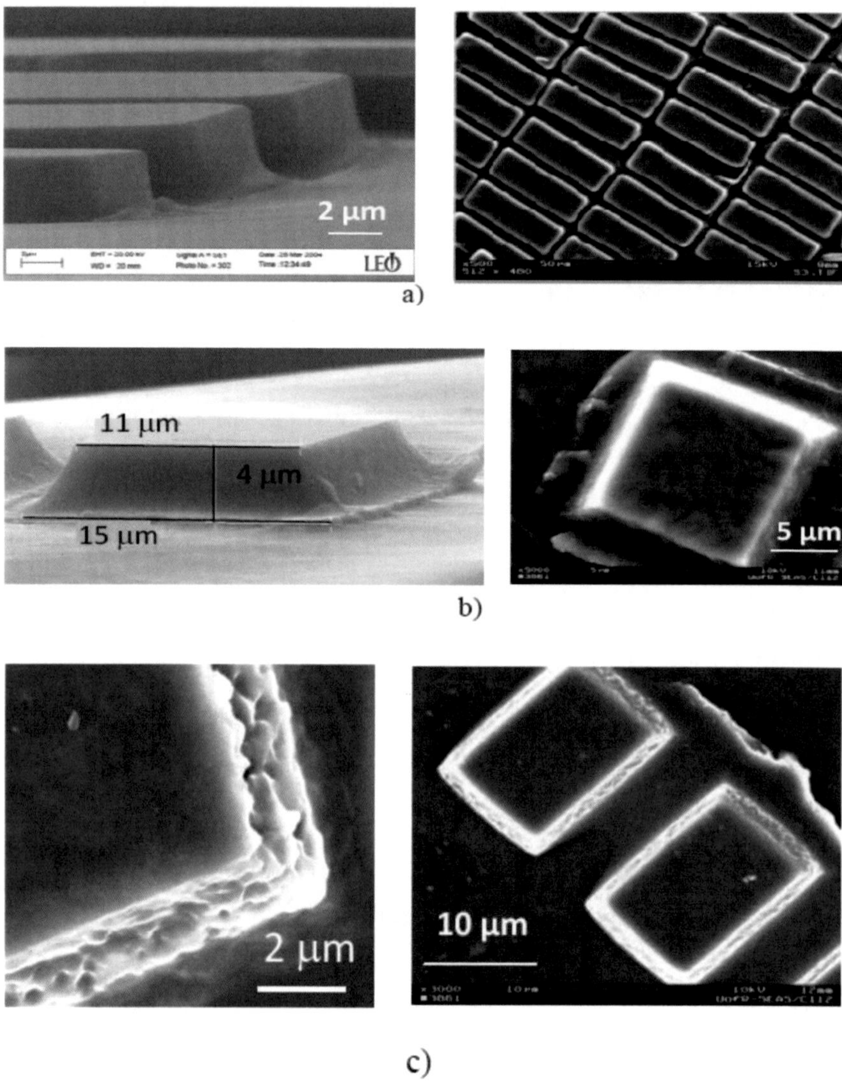

Figure 14. Shaped PCLC microparticles characterized by SEM: a) dimensional uniformity of rectangle PCLC microparticles, b) single filling to produce square shaped microparticle, c) multiple filling of rectangle shaped microparticles [17].

4. Applications of PCLC Microparticles

PCLC particles can be used in different applications, passive and active. Possible *passive applications* include use in the military as decals (e.g., for identifying "friend and foe"), in decorative/cosmetic inks and paints (e.g., car paint), biomedical applications, for document security, in stereoscopic printing, as waveplates, filters and polarizers. For *active applications* PCLC microparticles can be used in camouflage of military vehicles, smart windows, color filters, multi-color flexible displays, smart cards, electronic labels and billboard signs; and for these applications, an electro-optic control is required.

Also, PCLC microparticles have potential as an alternative to existing liquid crystal display (LCD) and particle display technologies. Due to their intrinsic cholesteric nature, PCLC microparticles can provide bright, saturated colors and circular polarization effects without needing additional color filters and polarizers. These advantageous properties of the microparticles provide a very good base for a new generation of particle displays like *electronic paper*: a flexible, thin, multi color, low-power, and low-cost display [2,13-16,27-37].

CONCLUSION

This work presents a way of manufacturing tailored PCLC microparticles by use of soft lithography technique. This method is shown to be the most efficient and easy way of shaping PCLC material into tiny PCLC microparticles. Namely, a flexible PDMS mold taken as replica of rigid silicon wafer template can produce plenty of PCLC particles in a variety of determined shapes, sizes and thickness. There are many advantages of regularly shaped PCLC microparticles over irregularly shaped PCLC particles such as controllable dimensions, uniform thickness, well-controlled electro-optic response of the shaped microparticles under applied external field that can be easier model and predict (for active application), possibility of making layered structure of PCLC microparticles with the same or different type of materials, adding dopants in PCLC material to change physical, optical, electrical properties, etc.

PCLC microparticles present a novel group of materials that can find many applications—passive and active ones, such as polarizers, notch filters, novel type of full color displays with low power consumption, sensors, smart windows, medical applications, smart paints and inks, etc.

ACKNOWLEDGMENTS

This work was done at the University of Rochester, Laboratory for Laser Energetics in Rochester, NY, USA. The author would like to thank Dr. Stephen D. Jacobs, Mr. Kenneth Marshall and Dr. Tanya Kosc for their supervision and guidance. The author also acknowledges the Laboratory for Laser Energetics for a generous Horton Fellowship.

REFERENCES

[1] Collings, P., Hird, P. *Introduction to Liquid Crystals: Chemistry and Physics;* Taylor & Francis Ltd.: London, UK, 1997.

[2] Yeh, P., Gu, C. *Optics of Liquid Crystal Displays;* Wiley & Sons: NY, 1999.

[3] Chandrasekhar, S. *Liquid Crystals;* Cambridge University Press: NY, 1977.

[4] Blinov, L. V. Electro-optical and Magneto-optical Properties of Liquid Crystals; Wiley & Sons Ltd.: NY, 1983.

[5] Tamaoki, N. *Adv. Mat.* 2001, *13*(15), 1135-1147.

[6] Berreman, D. W. *Phys. Rev. Lett.* 1970, *25*(9), 577-581.

[7] Binet, C., Mitov, M., Boudet, A. *Mol. Crys. Liq. Crys.* 2000, *339*, 111-123.

[8] Fergason, J. L. *Mol. Crys.* 1966, 1, 293-307.

[9] Chen, H. P., Katsis, D., Mastrangelo, J. C., Jacobs, S. D., Hood, P. *Adv. Mat.* 2000, *12*(17), 1283-1286.

[10] Kreuzer, F. H., Andrejewski, D., Haas, W., Haberle, N., Riepl, G., Spes, P. *Mol. Crys. Liq. Crys.* 1991, *199,* 345-377.

[11] Kreuzer, F. H., Haberle, N., Leigeber, H., Maurer, R., Stohrer, J., Weis, J. *Cyclic Liquid Crystalline Siloxanes: Chemistry and Applications.* Eds. Auner, N. Weis, J., *Organosililicon Chemistry III: From Molecules to Materials*; Wiley-VCH Verlag GmbH: Weinheim, Germany, 1996, 566-586.

[12] Chen, S. H., Chen, P. H. M., Katsis, D., Mastrangelo, J. C. *Chromogenic Phenomena in Polymers* 2005, *888*, 290-306.

[13] Korenic, E. M., Jacobs, S. D., Faris, S. M., Li, L. *Color Res. Appl.* 1998, *23*(4), 210-220.

[14] Schadt, M. *Annu. Rev. Mat. Sci.* 1997, *27*, 305-379.

[15] Schadt, M. *Mol. Crys. Liq. Crys.* 1988, *165,* 405-438.

[16] Korenic, E. M. *Colorimetry of Cholesteric Liquid Crystals,* PhD Thesis, University of Rochester, Rochester, NY, 1997.

[17] Trajkovska Petkoska, A. *Enhanced Electro-optical Behavior of PCLC Flakes in Host Fluids,* PhD Thesis, University of Rochester, Rochester, NY, 2007.

[18] Korenic, E. M., Jacobs, S. D., Faris, S. M., Li, L. *Mol. Crys. Liq. Crys.* 1998, *317*, 197-219.

[19] Faris, S. M. *Aligned Cholesteric Liquid Crystal Inks.* U.S. Patent No. 5,364,557, 1994.

[20] Kosc, T. Z. Motion of Polymer Cholesteric Liquid Crystal Flakes in an Electric Field, PhD Thesis, University of Rochester, Rochester, NY, 2003.

[21] Trajkovska Petkoska, A. Polymer Cholesteric Liquid Crystal Flakes – Their Electro Optic-Behaviour for Potential E-Paper Application; VDM Verlag Dr. Müller: Saarbrücken, Germany, 2008.

[22] Whitesides, G. M., Gates, B., Mayers, B., Xu, Q. B. *Abstracts of Papers of the American Chemical Society* 2005, *229,* U936.

[23] Whitesides, G. M. Abstracts of Papers of the American Chemical Society 1996, 212, Inor-031.

[24] Xia, Y. N., Whitesides, G. M. *Annual Review of Materials Science* 1998, *28,* 153-184.

[25] Trajkovska-Petkoska, A., Jacobs, S. D., Kosc, T. Z., Marshall, K. L. Manufacture of Regularly Shaped Polymer Cholesteric Liquid Crystal Flakes with a Mechanically Flexible Mold, U.S. Pat. 7,238,316 B2, 2007.

[26] Trajkovska-Petkoska, A., Varshneya, R., Kosc, T. Z., Marshall, K. L., Jacobs, S. D. *Adv. Funct. Mater.* 2004, *15*, 217.

[27] Pfaff, G., Reynders, P. *Chem. Rev.* 1999, *99*(7), 1963-1981.

[28] Wu, S.-T., Yang, D. K. *Reflective Liquid Crystal Displays*; Wiley & Sons Ltd.: NY, 2001.
[29] Doane, J. W. Polymer Dispersed Liquid Crystal Displays, Liquid Crystals, Applications and Uses; World Scientific: London, UK, 1990, vol. 1, 361-395.
[30] Penterman, R., Klink, S. L., de Koning, H., Nisato, G., Broer, D. J. *Nature* 2002, *417*(6884), 55-58.
[31] Comiskey, B., Albert, J. D., Yoshizawa, H., Jacobson, J. *Nature* 1998, *394*(6690), 253-255.
[32] Sheridon, N. K., Berkovitz, M. A. Proceedings of the Society of Information Displays (SID) 1977, *18*(3-4), 289-293.
[33] Trajkovska-Petkoska, A., Jacobs, S. D., Marshall, K. L., Kosc, T. Z. Electrically Actuated Doped Polymer Flakes and Electrically Addressable Optical Devices Using Suspensions of Doped Polymer Flakes in a Fluid Host U.S. 7,713,436 B1, 2010.
[34] Trajkovska-Petkoska, A. *Polymer cholesteric liquid crystal flakes as new candidates for display and sensor applications.* NATO Science for Peace and Security Series-B: Physics and Biophysics: Nanotechnological Basis for Advanced Sensors, Springer, 2011.
[35] Trajkovska-Petkoska, A., Jacobs, S. D. *J. Mat. Sci. Eng. (A&B),* A2 (2), 2012, 137-151.
[36] Kosc, T. Z., Marshall, K. L., Trajkovska-Petkoska, A., Kimball, E., Jacobs, S. D. *Displays.* 2004, *25*, 171–176.
[37] Li, Q. Nanoscience in Liquid Crystals: from self-organized nanostructures to applications; Springer: Heidelberg, 2014.

In: New Developments in Liquid Crystals Research
Editors: A. T. Petkoska and A. T. Broach
ISBN: 978-1-63482-456-9

Chapter 2

MODIFICATION OF ELECTRO-OPTICAL RESPONSE OF TAILORED PCLC MICROPARTICLES

Anka Trajkovska Petkoska*

University St. Kliment Ohridski-Bitola,
Faculty of Technology and Technical Sciences-Veles, R. Macedonia

ABSTRACT

Polymer Choelsteric Liquid Crystal Material (PCLC) made in a form of tailored microparticles have been proposed for novel optical applications. PCLC microparticles exhibit unique optical properties, such as selective reflection and circular polarization of reflected light. Proper molecular design can yield PCLC materials with varying pitch lengths that can reflect color from deep UV- to far IR- region. Preparation of PCLC films and irregularly-shaped PCLC particles have been already reported. Due to certain disadvantages of irregularly-shaped PCLC particles, it was of crucial importance to find a way to make tailored PCLC particles with predetermined shapes, sizes and properties. Details of the preparation method of shaped PCLC microparticles by soft lithography, their characterization, and possible applications have been reported by the author and her team.

In this chapter, modification of PCLC material/microparticles is described. Variety of dopants added to the PCLC material showed different effects. Dopants of different nature, *viz.* metal-based and carbon-based; conductive and non-conductive dopants, dopants with nano- and micron-sizes; dopants of different shapes such as spherical, elongated or flaked forms, are presented here. Of particular interest were the dopants that influence the dielectric properties of PCLC material. Further, electro-optical behavior of neat and doped PCLC microparticles is investigated under an applied electric field in AC and DC regime. This electro-optic behavior of PCLC particles opens novel applications, beside the known *passive applications*, such as: use in the military as decals, decorative/cosmetic inks and paints, document security, stereoscopic printing, retarders, waveplates, filters, polarisers, etc. In *active applications,* the orientation of PCLC microparticles can be controlled by use of an external electric field as smart windows,

* Corresponding Author address: Email: anka.trajkovska@uklo.edu.mk

color filters, multi-color flexible displays, such as "electronic paper", smart cards, electronic labels, various sensors, billboard signs and so on.

Keywords: Polymer cholesteric liquid crystals, shaped PCLC microparticles, electro-optical behaviour, dopants

1. Introduction

Polymer cholesteric liquid crystal (PCLC) materials have been used in forms of continuous films and flakes (microparticles). They exhibit unique optical properties, such as selective reflection and circular polarization due to their helical structure. Proper molecular design of LC molecules enables PCLC materials with varying pitch lengths that can reflect color from deep UV- to far IR- region, as well as ability to be processed easily into large-area, monodomain, defect-free films. The unique optical characteristics of PCLCs, such as the selective reflection and circular polarization, have potential in many optical, photonic and electronic applications [1-3].

PCLC microparticles have been described earlier by the author and her team. Initially, they were made by freeze-fracturing of PCLC continuous films with liquid nitrogen and irregularly shaped PCLC microparticles (flakes) are obtained. Due to variation in dimensions, shapes, thicknesses, these PCLC microparticles exhibit variation in their electro-optic behavior when they are placed in host fluids. Later, by use of soft lithography technique, tailored shapes of PCLC microparticles are manufactured. The tailored PCLC microparticles showed determined and uniform behavior in a variety of host fluids under an applied AC electric field. Further, PCLC material was modified by different types of dopants like, carbon-based dopants, metallic and inorganic dopants in order to modify the dielectic properties of polymer matrix. This enables to extend the use of PCLC microparticles not only in AC, but also in DC regime [4-10].

These PCLC mcroparticles have the potential to be used in numerous passive and active electro-optic applications. The most attractive application of PCLC microparticles is as an alternative to existing liquid crystal display (LCD) and particle display technologies, because they can provide number of advantages, such as *a flexible, thin, multi color, low-power, and low-cost display.* Also, they can be used in switchable color filters, optical retarders, and micropolarizers, photonics applications, vehicle camouflage, etc.[8-10].

2. Experimental

2.1. Neat PCLC Films and PCLC Microparticles

PCLC material used in this work is non-crosslinkable polysiloxane PCLC. The material possesses a low degree of polymerization (low molecular weight), which results in a relatively low viscosity. Usually, PCLCs don't crystallize and remain in glassy state; glass transition temperature (T_g) is about of ~ 50 ^{0}C and a clearing point (T_c) of ~ 200 ^{0}C. Figure 1 presents PCLC material in form of thin film (Figure 1a), irregularly (Figure 1b) and regularly shaped PCLC microparticles (Figure 1c) [5-8,11].

a) b) c)

Figure 1. a) PCLC film prepared on a silicon wafer substrate. b) Irregularly shaped PCLC microparticles. c) A composite photo showing red, green and blue PCLC microparticles in various shapes [9].

Initially, irregularly PCLC microparticles were made by freeze-fracturing of PCLC film with liquid nitrogen (e.g., Figure 1a and b). Due to the disadvantages of these microparticles, such as variying of shapes, sizes and thicknesses, it was found a way to make shaped forms of PCLC materials (Figure 1c). The irregular shaped PCLC microparticles showed random or non-reproducible electro-optic behaviour when they are placed in a host fluid and subjected under an external electric field [12].

Soft lithography was used as a technique for shaping PCLC material into tailored microparticles with determined shapes, sizes and thicknesses. For this purpose, a flexible elastomeric PDMS mold is created as a replica over a template that is a patterned silicon wafer. The PDMS mold was used to make a variety of shapes of PCLC microparticles [5-8].

Regularly shaped PCLC microparticles are observed in a variety of host fluids under an applied external electric field aimed for some active application. Their electro-optic (EO) behaviour was determined and reproducible. They could be used for electro-optical devices e.g., as reflective displays, e-book application, smart windows, that consumes low power energy. More about the electro-optic behaviour of PCLC microparticles follows in Section 3 [6,10,13].

2.2. Dopants

There are a variety of particulate substances, like carbon-based and metallic, that can be used as dopants (fillers, inclusions) in polymer materials forming composites with unique properties. Particle *dopants* can be of any shape (e.g., spherical, fiber-like, tubes, or even flake-platelets) and size (from nano- to micron-sizes). The final heterogeneous composite mixtures exhibit improvements over those of the polymer matrices in terms of mechanical, electrical, and thermal properties, tailoring them to different applications.

Different types of dopants were selected to modify the PCLC properties. Of particular interest were the dopants that affect the dielectric properties of PCLC material. The used dopants in this study varied in their sizes (nano- and micron- sized dopants), different dopants' shapes were used (spherical, elongated or flaked forms), as well as dopants of organic and inorganic nature were studied. Table 1 gives a brief description of the dopants studied in this work. They are categorized in three groups: [6,14]

1) *carbon-based dopants* (conductive, nano- and micro-sized): carbon black, carbon nanotubes;
2) *metal-based dopants* (micron-sized): carbonyl iron, ITO, Al flakes; and
3) *inorganic dopants* (non-conductive dopants, but with high dielectric constant, nano- and micron-sized): $BaTiO_3$, TiO_2.

Table 1. List of used dopant materials in this work

Dopant	Particle size*	Percolation threshold (PT) (vol %)	Conductivity (1 kHz) at PT [S/m]	Dielectric constant (10 kHz) at PT (vol %)
Neat PCLC (no dopant)	–	–	$\sim 1 \times 10^{-9}$	2.1
1) C-based dopants				
CB VPA	17 nm	2	1×10^{-5}	10
CB M	75 nm	6	1×10^{-5}	10
SWNT-1	1.4 × 2 to 5 μm	0.7	1×10^{-6}	5
SWNT-2	1.1 nm × 0.5 to 100 μm	0.4	1×10^{-6}	2.2
MWNT	10 to 20 nm × 0.5 to 200 μm	0.8	$\sim 1 \times 10^{-3}$	10
2) Metal-based dopants				
CI-HQ	1.1 μm	20	1	–
ITO	3.38 μm	25	$\sim 1 \times 10^{-4}$	5
Al flakes	12 μm	30	$\sim 1 \times 10^{-2}$	20
3) Inorganic dopants				
BT-1	70 nm	–	–	7.7 (27.0)
BT-2	1 to 1.1 μm	–	–	6.1 (32.0)
BT-3	1.3 to 1.8 μm	–	–	5.8 (27.0)
TiO_2-1	30 nm	–	–	5.2 (54.4)
TiO_2-2	10 nm × 40 nm	–	–	7.5 (22.0)

*Primary average particle dimensions.
Legend: CB VPA – carbon black Vulcan PA90, CB M – carbon black Monarch M120, SWNT-1 is single carbon walled nanotubes (type 1), SWNT-2 is single walled carbon nanotubes (type 2), MWNT – multiwalled carbon nanotubes, CI-HQ – carbonyl iron, ITO – indium tin oxide, BT – barium titanate.

Conductive dopants like, carbon-based or metallic dopants can convert an insulating polymer to an electrically conductive material. The insulating-to-conducting transition of the polymer composition occurs when the dopant concentration exceeds a critical value known as the *percolation threshold* (PT). PT is characterized by a sharp jump in the conductivity by several orders of magnitude. The location of the PT depends on many factors: polymer host (viscosity, surface tension and type), the dopant nature (size, shape, geometry, surface energy), the composite porosity, dopant dispersion and alignment, chemical interactions, processing method, presence of surfactants, etc. Generally, it is desired to have smaller PT values. It could be achieved with elongated dopants compared to spherical ones. The aspect ratio (length : diameter) of dopants affects the PT; the higher the dopant aspect ratio, the lower the PT. Better dopant dispersion also contributes to lowering PT. But, good alignment (e.g., nanotubes in polymer matrix) does not always give a lower PT; there are fewer contacts between the aligned nanotubes, resulting in reduction of electrical conductivity and consequently higher PT. Also, higher PT is expected in the composites where stronger dopant/polymer interactions occur (due to the slower creation of conductive network). An increase in surface tension of the polymer can increase the PT, because this prevents the formation of conductive particle clusters [15-37].

There are different *percolation models* which interpret the behavior of the electrical conductivity of different mixtures: statistical, geometrical, thermodynamic and structure-oriented. The classical statistical model explanation of the *conductivity of a composite (mixture),* σ_m, is due to the existence of "infinite clusters" of connected conducting particles (forming a three dimensional conductive network over the entire volume of the sample) and is given with a power-law dependence:

$$\sigma_m = \sigma_0 (p - p_c)^t \quad \text{or} \quad \sigma_m \approx (p - p_c)^t \quad \text{for } p>p_c$$

where: σ_0 is conductivity of the filler's particles (or fitting constant), p is the volume fraction of the filler, p_c is the filler's critical volume fraction at the percolation threshold ($p_c \cong$ PT), and t is the conductivity exponent determining the increase of the conductivity above the PT (usually $t = 1.5$-2).

The *dielectric constant of the composite,* ε_m, also exhibits a power-law behavior:

$$\varepsilon_m = c\left(\frac{p_c}{p - p_c}\right)^s \quad \text{or simplified form is:} \quad \varepsilon_m \approx (p_c - p)^{-s} \text{ for } p<p_c$$

where: c is a constant and s is the dielectric permittivity critical exponent (usually reported to have values of $|s| = 0.7$-0.9, absolute values, but also departures from the universal values for t and s have been reported). For example, the value of t can be $t = 1$ to 7 for elongated particles or dopant aggregates (non-universal percolation) or even much smaller $t < 1$ [6].

Statistical (classical) theory gives a good description of PT. The critical volume fraction for percolation is calculated to be from 5 to 20 vol.% (sometimes even higher than 50 vol.%).

However, discrepancies are observed between critical parameters like p_c, t, s and the experimental results. This is due to the fact that the classical theory does not cover the properties of the polymer host (viscosity, surface tension), polymer/dopant chemical interactions and dopant geometry. Physical or "electrical" contacts between spherical particles are very important in a description of classical percolation theory. "Continuous chains" of particles do not always exist in conductive composites as the theory assumes. Discontinuities (e.g., thin polymer barrier layers) may occur between the particles. In this case, the composites can exhibit a semiconducting or conducting character with conduction occurring through a "tunneling" effect or a "hopping" of charge-carriers among neighboring particles. Consequently, much lower percolation thresholds are observed (lower than 0.1 vol.%), than the classical theory would predict. There are currently no better theories for predicting the full range of experimental results observed [6].

Usually, the percolation threshold (PT) value can be calculated with the following equation:

$$PT \cong p_c = \frac{1}{1+\rho(DBP)\frac{C_{\max}}{C_{crit}}}$$

where: ρ is the density of the conductive dopant, DBP is the oil absorption (absorption of dibutyl phthalate-DBP oil) on the dopant [cm^3/g], C_{crit} is the number of contacts of a particle with neighboring particles at the critical concentration (at PT), C_{max} is the maximum number of contacts which the particle can have with its neighbors.

Inorganic dopants, like $BaTiO_3$ and TiO_2 are considered as *non-conductive* dopants, but with high dielectric constants. They increase the dielectric constant of PCLC when they are added in large concentrations, but conductivity is not affected. For these polymer compositions it is useful to be able to predict/calculate the dielectric constant as a result of doping. Various dielectric mixing models for prediction of the dielectric constant of heterogeneous mixtures are known. For example, the equations for the effective dielectric constant of the composition are based on the dielectric permittivity of each constituent and its volume fraction. The models and equations given below, have been already described in literature, and are used for the purpose of this work [42-45].

Some authors cite a *Maxwell-Garnett*-like mixture equation:

$$\frac{\varepsilon_m - 1}{\varepsilon_m + 2} = f_1 \frac{\varepsilon_1 - 1}{\varepsilon_1 + 2} + f_2 \frac{\varepsilon_2 - 1}{\varepsilon_2 + 2}$$

where ε_m is the dielectric constant of the mixture, ε_1, ε_2, f_1 and f_2 are the dielectric constants and volume fractions of each component (1 refers to the host matrix or PCLC, and 2 for dopant or filler. Models for the effective dielectric constant of composites use the same notation.)

The *Bruggeman* (*Botcher* equation or *Bruggeman-Hanai*) equation is given as:

$$f_1 \frac{\varepsilon_1 - \varepsilon_m}{\varepsilon_1 + 2\varepsilon_m} + f_2 \frac{\varepsilon_2 - \varepsilon_m}{\varepsilon_2 + 2\varepsilon_m} = 0$$

The *Logarithmic* equation (or *Lichtenecker-Rother* eq.) is one of the most commonly used equations for predicting the dielectric constant of statistical/chaotic composites. It is given as:

$$\log \varepsilon_m = f_1 \log \varepsilon_1 + f_2 \log \varepsilon_2$$

For *insulating fillers* in a polymer, there is a Maxwell-Wagner mixing rule formula for calculating the dielectric constant, given as:

$$\varepsilon_m = \varepsilon_1 \left\{ \frac{[2\varepsilon_1 + \varepsilon_2 + 2f_2(\varepsilon_2 - \varepsilon_1)]}{[2\varepsilon_1 + \varepsilon_2 - 2f_2(\varepsilon_2 - \varepsilon_1)]} \right\}$$

The above equations give similar results for lower dopant loading (few %) in a polymer matrix. For higher volume fractions of dopant in a polymer, particularly in PCLC as well, there are often discrepancies between these models and experiments.

This is possibly due to: i) the high volume fraction of the dopant may increase the porosity of the composite material, or ii) the high dopant fraction may lead to agglomeration of the powder and a non-uniform distribution through the composition [6].

2.3. Modified PCLC Material

The goal of doping of PCLC was mainly to enhance the dielectric properties of this material. Variety of dopants to the PCLC material showed different effect (Table 1 lists dopants used in this work and their effect on PCLC matrix) [6,46].

Preparation of doped disks (plates) made from PCLC material and a particular dopant were prepared initially in order to determine the dielectric properties of the composite. For each system (PCLC/dopant), ~10 or more samples with different concentrations are prepared to be ~2 cm in diameter and ~0.5 mm in thickness. A "thin film-maker tool" with controlled heating and pressure is used. [Constant-thickness filmmaker for polymers consists of heated platens with temperature control and an air E-Z press, 20-ton hydraulic press.] First, target compositions with specified amounts of PCLC's and dopant material were measured, ground by mortar and pestle, and mixed well in a small vibrational capsule. The mixed powders are melted to 130°C under a pressure of ~ 4000 kg for ~ 5 min and then cooled down to room temperature. The disks (PCLC/dopant) are placed in an impedance/gain phase analyzer for measurement of dielectric characteristics of the doped samples. The output of this instrument is resistance R and capacitance C as a function of frequency. From these data, the conductivity and dielectric constant for each dopant/PCLC composition is calculated. These data are important in prediction and modeling of electro-optical behaviour of PCLC microparticles [6].

2.4. Doped PCLC Microparticles

Tailored PCLC microparticles made of doped PCLC materials (explained in previous Section 2.3) are very attractive in terms of their specific electro-optic behaviour in a variety of host fludis under an applied electric field even in DC regime. They are manufactured by use of a soft lithography technique, already described by the author [5,6].

The EO behavior of doped PCLC particles is different than the behaviour of neat shaped PCLC microparticles. Neat PCLC microparticles (that are with low conductivity and low dielectric constant) could not be used in low conductivity low dielectric constant host fluids like silicone oils, but only in more conductive hosts or hosts with relatively high dielectric constants.

Doped PCLC microparticles made of dielectrically enhanced PCLC material can be hosted in fluids like insulating silicone oils and show reorientational and translational electro-optic behavior (that has not been observed earlier with PCLC microparticles; it is explained later in Section 3) [6,46].

Figures below show some examples of doped PCLC microparticles. Namely, Figure 2 presents SEM (scanning electron microscopy) micrographs of 5 vol.% CB VPA-doped PCLC microparticles, while in Figure 3 are shown POM (polarizing optical microscopy) images of gradually doped PCLC compositions with carbon black CB M.

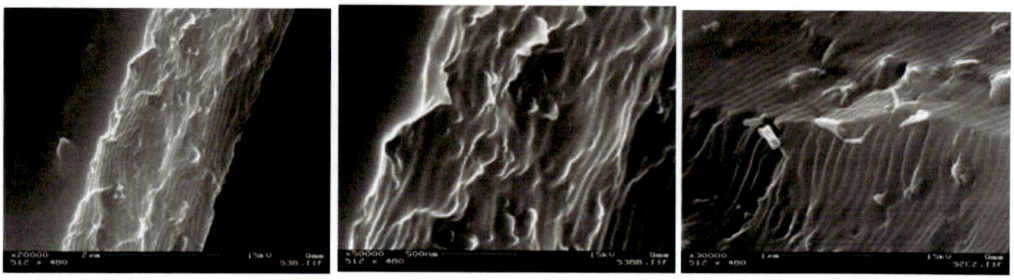

Figure 2. SEM pictures of 5 vol.% CB VPA dopant in PCLC.

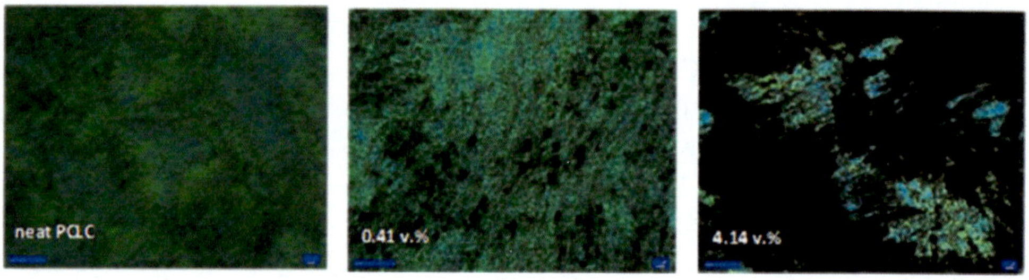

a) b) c)

Figure 3. POM micrographs of: a) neat PCLC, b) 0,41 vol.% and c) 4,14 vol.% CB M-doped PCLC.

Figure 4 shows PCLC microparticles doped with single walled carbon nanotubes (designated as SWNT-1, Table 1) in concentration of 1,2 vol.%.

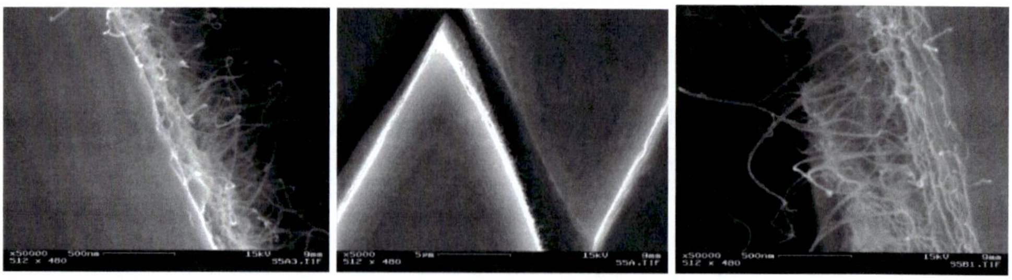

Figure 4. SWNT-1/PCLC composites: 1,2 vol.% SWNT-1 doped PCLC microparticles.

Figure 5 shows SEM and POM images of $BaTiO_3$-doped PCLC microparticles, while POM images of TiO_2-doped PCLC microparticles are presented in Figure 6.

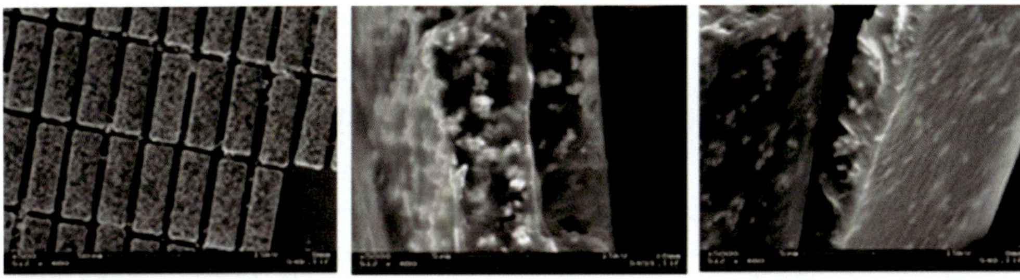

a)

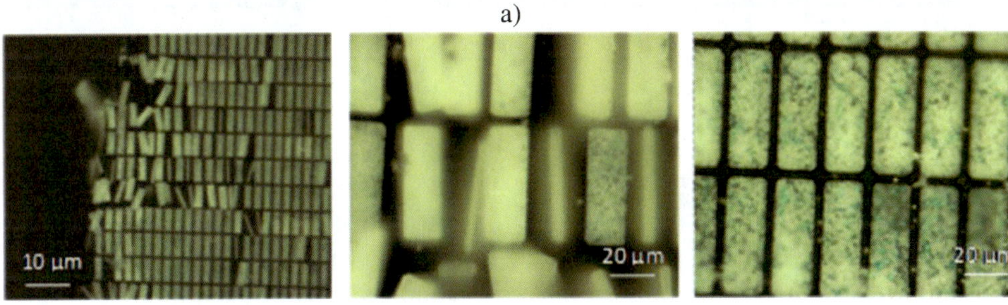

b)

Figure 5. a) SEM images, and b) POM images of 19.5 vol.% BT-1 doped shaped PCLC microparticles.

Figure 6. POM images of highly TiO_2-doped PCLC microparticles.

2.5. Host Fluids

Host fluids used in this work can be categorized in two groups: (i) hosts that works for PCLC microparticle system in AC regime, and (ii) hosts for DC regime. Also, there is a group of hosts which can be used as intermediates, and they work in DC as well as in a low-frequency AC field [6].

Compatibility of the host fluid with PCLC microparticles is very important for a good selection of the host fluid. Other properties like, host conductivity, dielectric constant, viscosity, vapor pressure, density, are also important and they have to be considered in selection of the proper system for electro-optic application of PCLC microparticles.

The potential *host fluids for AC regime* are given in Table 2. They are as follows: gamma butyrolactone (GBL), propylene carbonate (PC), ethylene glycol (EG), methoxy polyethylene glycol (MPEG350) and polyethylene glycol 200 (PEG200). Table 2 also gives some of important properties for these colorless hosts. The most preferable hosts are those with a higher dielectric constant and a lower viscosity (e.g., hosts like, GBL and PC).

Table 2. Properties of host fluids used in the AC regime (ordered according to increasing viscosity)

Host fluid	η_0 [mPa.s] at 25^{0}C	ρ_h [kg/m^3] at 20^{0}C	p_v [mmHg]† at 20^{0}C	n_r at 20^{0}C (λ=589.6nm)	ε_h at 100 kHz: measured/literature value	σ_h at 1 kHz [S/m]
GBL	1.7	1144	1.5	1.436	40 / 40	4.5 x 10^{-5}
PC	2.0	1200	0.03	1.421	60 / 65	7.5 x 10^{-4}
EG	15	1116	0.06	1.431	37 / 40	4 x 10^{-5}
MPEG 350	28	/	0.05	/	12 / NA	2 x 10^{-5}
PEG200	55	1083	0.01	1.463	18 / 17	8 x 10^{-5}

† p_v vapor pressure, conductivity (σ_h), dielectric constant ($\boldsymbol{\varepsilon_h}$).

Host fluids used in DC regime consist of different types of silicone oils. Table 3 compares some important properties of silicone oils considered for use in our work. The oils are of different chemical compositions and colorless. The DMS T-types are polydimethylsiloxane trimethylsiloxy terminated (the last number on the name means the oil viscosity), DMS E-series are epoxypropoxypropyl terminated polydimethylsiloxane oil. SIT 7757 is a 1,1,5,5-tetraphenyl-1,3,3,5-tetramethylsiloxane. Most of the oils shown in the table possess low conductivities (σ_h) and low dielectric constants ($\boldsymbol{\varepsilon_h}$) [6].

An advantage of these oils is the fact taht they do not change their conductivity over a longer period of time, unlike the AC hosts that are hygroscopic. An optimal host would have a low viscosity and low vapor pressure (p_v), as well as a close density match to PCLC particles. DMS T00 and T01 evaporates very fast from a test device, even if the viscosity is the most preferable of all DC hosts in the table. T05 oil has a much lower vapor pressure than T00 or T01, and it is much easier to work with it in the test devices over long periods of time.

Table 3. Properties of some silicone oils (ordered according to increasing viscosity)

Silicone oil	η_0 [mPa.s] at 25^{0}C	ρ_h [kg/m^3] at 20^{0}C	p_v [mm Hg] at 20^{0}C	n_r at 20^{0}C (λ=589.6 nm)	ε_h at 100 kHz (meas)	σ_h at 1 kHz [S/m]
DMS T00	0.5	764	55	1.375	2.18	1 x 10^{-8}
DMS T01	0.82	820	4	1.382	2.10	6 x 10^{-9}
DMS E01	1	995	NA	NA	6.60	4 x 10^{-7}
DMS T05	4.6	918	NA	1.397	2.55	6.6 x 10^{-9}
DMS E09	8-11	995	NA	1.446	6.60	3 x 10^{-8}
SIT 7757	35-40	1070	<0.001	1.551	2.70	3 x 10^{-8}
PDM 7050	170	1090	<0.001	1.588	~ 2.5	~ 1.10^{-9}

Neat PCLC flakes in the low conductivity, low dielectric constant oils do not reorient to 90^0 when the electric field is applied. At best they exhibit non-uniform flake motion for large applied electric fields. In E01 oil, neat PCLC flakes reorient to 90^0. This is because, for E01, there are large differences between ε_p and ε_h, as well as between σ_p and σ_h. The E09 oil exhibits similar properties to E01 as a host, but with a higher viscosity.

2.6. Electro-Optic Device for Testing PCLC Microparticles

An electro-optic device based on PCLC microparticles consists of a cell assembled of two indium tin oxide (ITO)-coated glass substrates, which is filled with a suspension containing PCLC microparticles in a suitable host fluid. Usually, ~1 wt.% PCLC microparticles were tested in a particular host fluid in order to study single particle motion (and not influenced by neighboring particles). Figure 7 presents such electro-optical device based on PCLC particles that can be used as a reflective type display, e.g., particle display for electronic paper application [6,9,12].

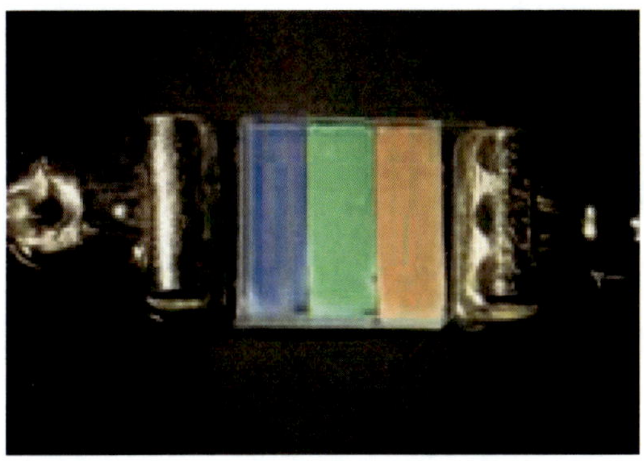

Figure 7. Test device for electro-optic behaviour of PCLC microparticles: 3-segmental device filled with blue, green and red PCLC microparticles in suitable host fluid.

3. RESULTS AND DISCUSSIONS

3.1. Effect of Dopants on PCLC Host

Carbon-based and metal-based dopants are conductive dopants. They convert PCLC material from an insulating polymer to an electrically conductive material. Actually, they show an increase of the conductivity of the composite by several orders of magnitude when the PT is reached. Percolation threshold (PT) was determined by measuring σ and ε as a function of doping level for a series of PCLC doped disks and those values are reported in Table 1 along with the used dopant materials.

There are two types of behavior for dopants in PCLC matrix. For carbon-based particles, PT drops as shape anisotropy increases (Figures 8a and b). Namely, in Figure 8a) two different types of carbon blacks used as dopants (CB VPA90 and CB M120) are compared in terms of conductivity values measured for series of PCLC/dopant compositions, as well as their PTs are designated in the graph (PT values are also reported in Table 1). Figure 8b) compares effect of different carbon nanotubes dopants (SWNT-1, SWNT-2, MWNT) on conductivity of different PCLC/dopant compositions. For metal-based particles with little shape anisotropy, PT increases with increasing particle size (CI-HQ, ITO, Al flakes) - Figure 8c) [6,46].

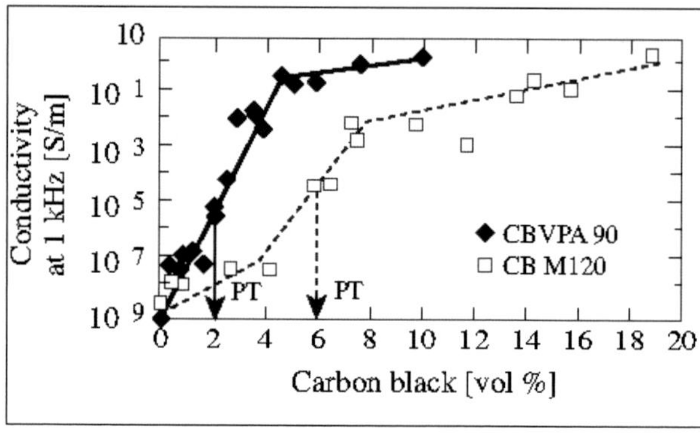

a)

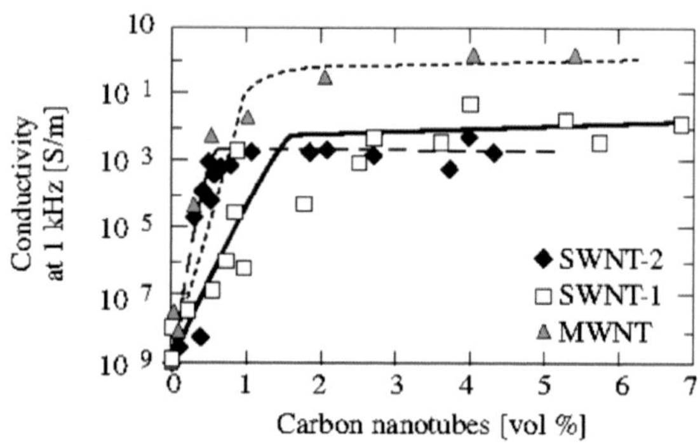

b)

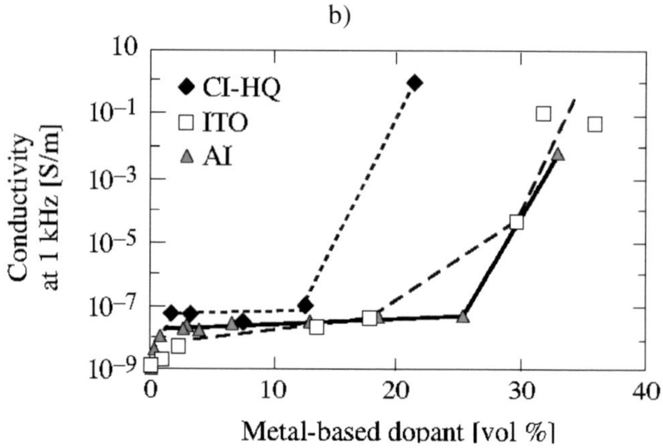

c)

Figure 8. Comparison of conductivity for PCLC doped with: a) carbon black dopants and their percolation thresholds, b) carbon nanotubes as dopants, and c) metal-based dopants. (Lines are drawn to guide the eye).

Figure 9 compares all of conductive dopants in PCLC material that are used in this work. Figure 9a) presents conductivity data for all used carbon-based dopants in PCLC composition (carbon blacks and carbon nanotubes). The elongated conductive dopants (SWNTs, MWNT) show an increase in conductivity of PCLC composition at lower dopant concentration compared to not elongated dopants such as carbon black dopants (CB M and CB VPA) in the same PCLC host material. Figure 9b) presents PT vs. aspect ratio of the dopants or dopant size (diameter).

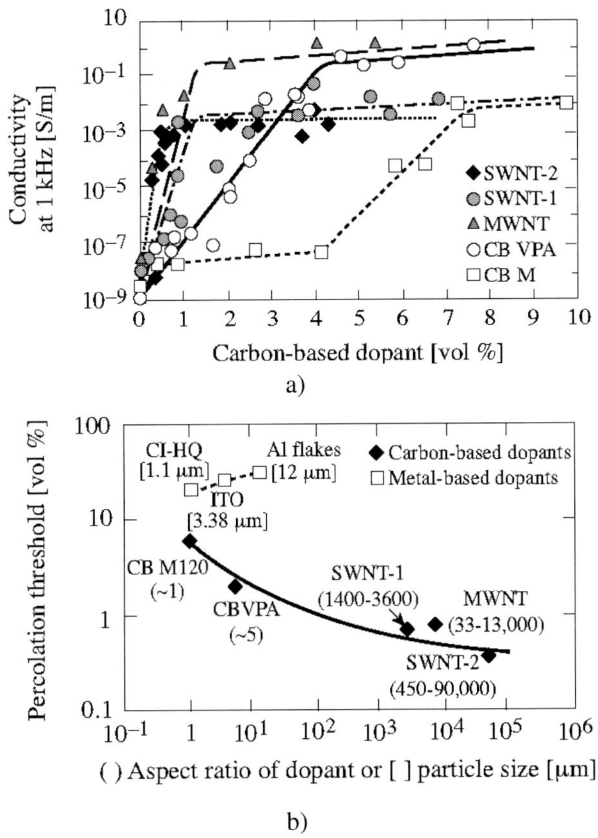

Figure 9. (a) Carbon-based dopants: comparison of conductivity at 1 kHz versus dopant concentration, (b) PT versus dopant average aspect ratio (range given in parentheses) or particle size (given in brackets). (Lines are drawn to guide the eye).

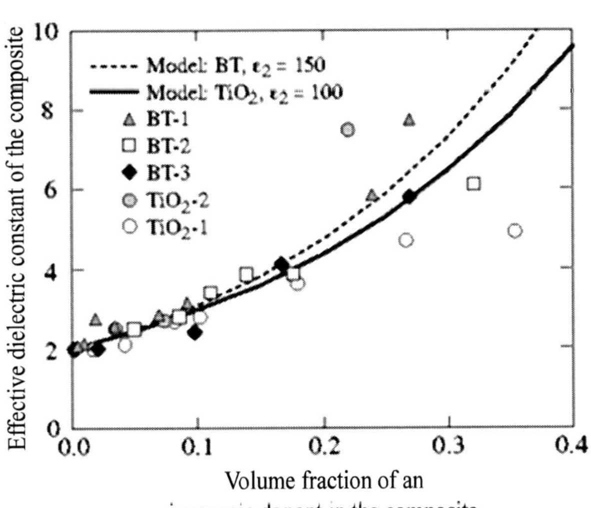

Figure 10. Dielectric constant at 10 kHz for PCLC composites using different inorganic dopants: $BaTiO_3$ and TiO_2, measured vs. calculated data.

In the case of carbon-based dopants, the most elongated dopants caused lower PT in PCLC compositions, while for metal-based dopants smaller particle sizes contributes for lower PT in the polymer composition.

The other class of dopants, *inorganic insulating dopants*, change the dielectric constant from 2 (for neat PCLC) to <10 at dopant levels of ~30 vol %, accompanied by an increase in density. [A logarithmic equation for the effective dielectric constant of the composite agrees well with our experimental data.]

Figure 10 compares all inorganic dopants that are not conductive, but have high bulk dielectric constants. An effective dielectric constant of PCLC composites vs. volume fraction of inorganic dopant is presented; measured vs. calculated (with logarithmic equation) values of dielectric constants are presented in this figure.

3.2. Electro-Optical Behavior of PCLC Microparticles

PCLC particles suspended in a host fluid reorient in an electric field due to Maxwell–Wagner (interfacial) polarization; charges accumulate on the particle surface, induces a dipole moment. The applied electric field acts on this induced dipole moment and causes the particle to reorient to a more energy stable configuration. Particle reorientation due to interfacial polarization performs most efficiently in AC electric fields of a specific frequency, and this frequency is determined by the dielectric properties of both the particle and the host fluid. Since PCLC microparticle are highly dielectric, the dominant parameter in the suspension is the host fluid conductivity. The electro-optic (EO) reorientation of the microparticles depends on many factors: host viscosity and its dielectric properties, as well as particle shape, thickness, and dielectric properties of the PCLC material. The applied voltage and its frequency are used to probe the particle reorientation, which could be faster, slower or non-existent. Usually, reorientation happens along the longest axis of the particle.

Particle reorientation in AC and DC regimes is described briefly in the following sections. Particle reorientation as a function of frequency, host viscosity, applied voltage, PCLC material nature, dopants, are determined experimentally and already reported by the author [6,12,13,47].

3.2.1. AC Regime

When an alternating current (AC) field is applied to the test device (Figure 11a), PCLC microparticles that initially lie nearly parallel to the cell substrates rotate for 90^0 about their longest axis. For example, an applied electric field of $\geq$ 10 $mV_{rms}/\mu m$ (*rms*: root mean square) between 10 and 1000 Hz, initiates particle reorientation with temporal response on the order of ten to hundreds of milliseconds. Selective reflection colors are shifted and diminished as the particles rotate (Figure 11b). Figure 11c) shows inducing a dipole on a PCLC particle when an electric field is applied [6, 12,13].

Figure 12 shows an example of electro-optic switching of square shaped PCLC particles in host fluid. Namely, square shaped particles of 40 μm on a side in PC host before and after application of 2.82 V_{rms} (100 Hz) across an 80 μm thick cell gap (or it is 35.2 $mV_{rms}/\mu m$).

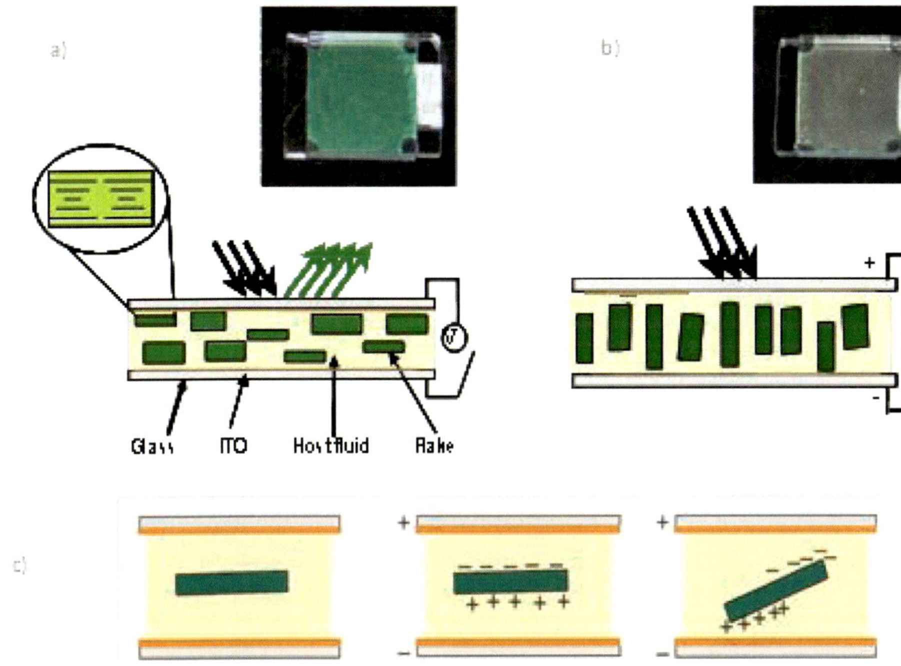

Figure 11. Typical cell device filled with a suspension of PCLC microparticles and host fluid: a) before, and b) after application of an electric field. *Insets*: photos of actual cell devices (top view): before (reflective-bright state) and after application of the electric field (dark-nonreflective state); c) Generation of a dipole moment on a flake particle when an electric field is applied (Maxwell-Wagner polarization).

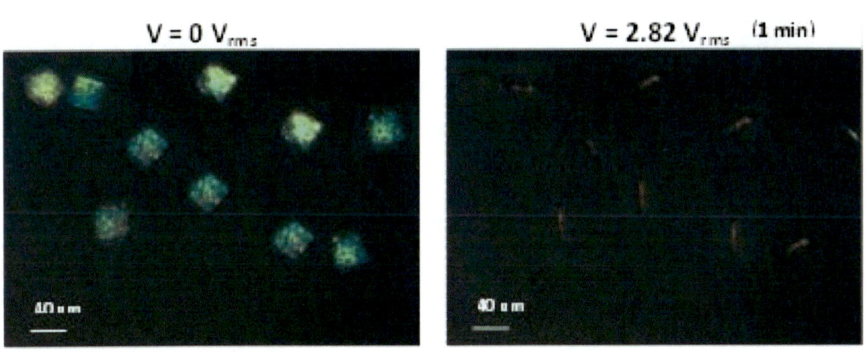

Figure 12. POM images of square microparticles in PC in test EO cell: a) before (reflective, bright state) and b) after reorientation (dark state) when the electric field is applied to the test cell device [6].

All particles reorient in ~ 1 min. [*Reorientation time* for the particles in a given host fluid is measured from the moment when an electric field is applied to the cell until flakes reorient to 90^0 (from horizontal to vertical position).] The motion is observed/measured under the microscope coupled with a camera/software [6].

Doped PCLC particles with increased conductivity in host fluids like, PC, show the same motion of reorientation as neat PCLC particles in the same host- PC. They reorient along the longest axis. *PCLC particles doped with high dielectric constant dopants*, like TiO_2 -1 (~20 vol.%) have a modestly increased dielectric constant (ε_p ~ 5). These doped particles show faster reorientation times (e.g., 0.7 s at 50 Hz) than neat PCLC particles [6].

3.2.2. DC Regime

The DC regime is of particular interest, because the use of low conductivity host fluids has benefits compared to the use of higher conductivity AC hosts. DC hosts do not exhibit large changes in conductivity with time, and they are less expensive (DC host fluids are listed in Table 3). Low viscosity hosts were preferred for use in our work, because faster reorientation times are expected, e.g., silicone oils like, SIT 7757, DMS T05 and DMS E01/E09.

Neat PCLC microparticles do not show motion (reorientation) in low conductivity, low dielectric constant host fluids (silicone oils) like, DMS T05 or SIT 7757, until relatively high voltages are applied or when very high particle concentrations are used (e.g., ≥ 10 wt.% PCLC particles in suspension, that could provide enough ions, charges in the system). In these cases, the motion is not uniform from particle to particle, motion is chaotic.

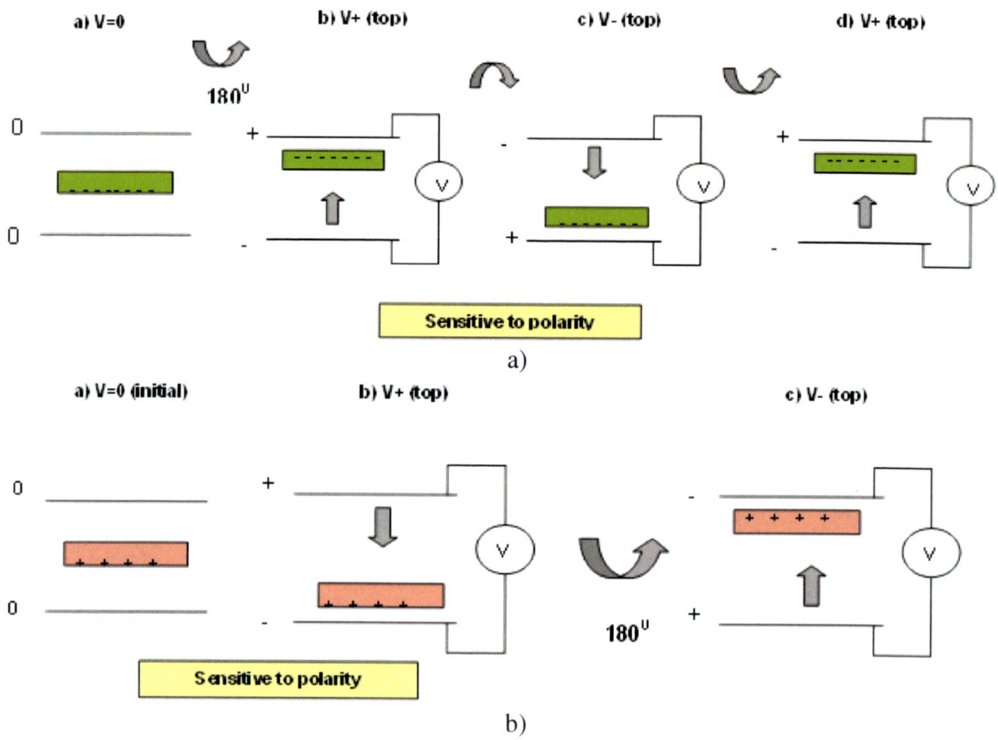

Figure 13. Schematic illustration of: a) CB VPA-doped PCLC particle and its EO behavior when positive, negative and again positive polarity is applied to the top electrode, b) SWNT-2 doped PCLC particles in DC host fluid when positive and negative polarity is applied on the top electrode [6].

PCLC microparticles with increased conductivity (PCLC particles doped with conductive dopants, Table 1) were tested in some of the considered oils, such as: DMS T05, SIT 7757. The purpose of doping is to increase the difference in dielectric properties between the host and the particles (particularly, increase σ_p). In DMS T05 and SIT 7757 hosts, these PCLC microparticles behave the same: translational motion (up to the top of the cell, and down to the bottom of the cell) depending on the polarity of the applied field, plus 180^0 reorientation [6,13].

Experiments on *particles doped with conductive dopants*, such as: CB or SWNT, showed that they possess a charge, a negative or a positive (concluded by measuring of zeta potential of doped PCLC material). This is confirmed by applying a positive/negative polarity to the top ITO electrode for a cell resting on the support stage in the microscope, and noting that the charged particles move up or down depending on the particle charge and applied polarity on the top electrode. Switching the polarity of the top electrode in the test device causes the particles to move up-down presenting bright and dark states (Figure 13).

PCLC microparticles with increased dielectric constant (e.g., doped with TiO_2) show 90^0 reorientation in low σ_h and low ε_h silicone oils, like DMS T05. The particles reorient along the longest axis, but also reorientation along the shorter particle axis is sometimes observed. This could be due to the nonuniform distribution of the dopant through the particle volume, or due to the variations in the particle thickness [6,13].

3.2.3. Summary of Electro-Optic Behaviour of PCLC Microparticles

Electro-optic behavior of PCLC microparticles is observed when driven with AC or DC fields, and they relax from a reoriented state when the field is removed. Metrics used to characterize the particle behavior are the frequency range over which response is observed and the response time in the AC regime. In the DC regime different types of reorientation and/or translation are observed, depending on the nature of the particles (neat or doped) and the host (low or moderate host, depending on ε_h and σ_h) [6,13].

The general observations for EO behavior of PCLC microparticles could be summarized as follows:

1) for AC regime:
   a) Particles in low viscosity fluids respond fastest,
   b) Particles in more conductive hosts respond at higher frequencies,
   c) The voltage dependence of the particle reorientation time confirms Maxwel-Wagner mechanism,
   d) Higher aspect ratio particles respond most quickly,
   e) Thinner particles respond faster than thicker ones,
   f) Particle concentration dependence on reorientation time is observed-at relatively high particle concentrations, the reorientation times increases,
   g) Doping to increase σ_p reduces the particle reorientation time,
   h) Doping to increase ε_p reduces the particle reorientation time,
   i) Higher conductivity ITO cells show faster particle response.

2) for DC regime:
   a) No reorientation occurs for neat PCLC particles in low σ_h and low ε_h hosts until a relatively high voltage is applied or high particle concentration is used in the test cell,
   b) Doped particles with increased σ_p reorient to 180^0 and exhibit a translational motion in low σ_h low ε_h hosts,

c) Doped particles with increased ε_p reorient to 90^0 in low σ_h low ε_h hosts,
d) Neat/doped particles show 90^0 reorientation in low σ_h and moderate ε_h hosts.

3.3. Application of Electro-Optic Devices Based on PCLC Microparticles

Some applications of PCLC particles (neat, doped or composite ones) in EO devices are presented in this paragraph. Figure 14 presents a possible *display application* of PCLC microparticles. Namely, Figure 14a) shows a composite PCLC particle consists of a layer of neat green PCLC and a layer of red SWNT-doped PCLC placed (e.g., positively charged) in a clear host fluid; switching the polarity of top electrode different reflected color could be observed (green vs. red).

Figure 14b) presents a device where the EO system consists of: two types of doped PCLC particles modified by conductive dopants resulting in different net charge of PCLC material. Depending on the applied polarity on the top electrode of the device, different reflected color from the PCLC particle could be observed (green vs. red) [6,8,10,13,14,47].

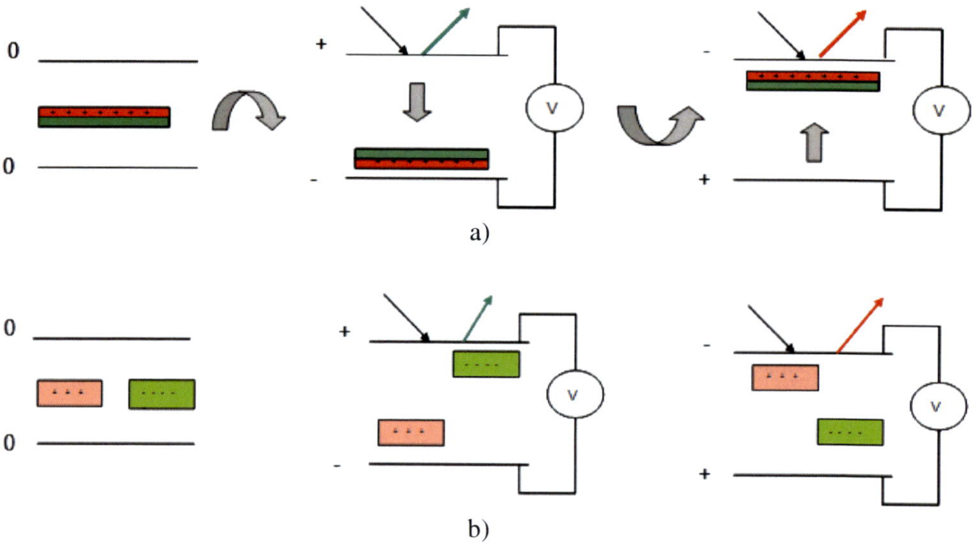

Figure 14. Some possible applications of doped PCLC particle-based device: a) composite particles consist of neat and doped layer in a clear host fluid, b) two different colored doped particles (single layers) in a dark host.

Also, devices based on PCLC particles could be used for *sensor applications,* for example they can work with visible light switching (bright-dark states) by using voltage-coded particles, e.g., particles with different shapes/sizes or particles made up of different PCLC materials that respond to different applied voltages. Also, chemical vapor sensors based on shaped PCLC particles are envisioned, e.g., particles that change the pitch length after exposing to certain chemical vapors; the vapors may cause "swelling" of the PCLC material, and thus, changes in the helical twist of the nematic sublayers, which in turn means shift in the particles' reflected color (example is given in Figure 15) [10].

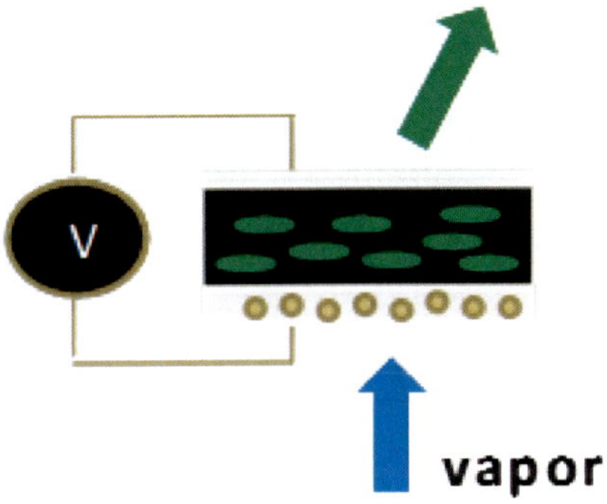

Figure 15. Vapour-sensitive device based on PCLC microparticles (Note: the substrates of the cell are vapour porous) [10].

CONCLUSION

PCLC particle technology holds promise as an alternative display technology to today's leaders in flexible reflective particle displays. This technology has some unique characteristics that no other particle technologies can provide, such as: each PCLC particle reflects a bright and saturated color, the reflected light is circularly polarized, and therefore, there is no need for additional color filters and polarizers.

The ability to create PCLC particles of a specific size and shape is an important development towards novel PCLC electro-optical devices. Particles with uniform shapes and sizes behave in the same manner: their EO behavior could be predicted and modeled. This uniformity can produce a crisp change in the device reflectivity. It is now possible to tailor the particle shape to provide the optimal reorientation time for a specific host fluid system. Actually, PCLC microparticles in an appropriate host fluid reorient from a bright (reflective) state to a dark (nonreflective) state under an applied AC electric field. Maxwell–Wagner polarization is confirmed as the main mechanism for the reorientation of these microparticles; or the electro-optic effect is created by extrinsic charges that migrate to the surface of the dielectric particle when an electric field is actuated, inducing a dipole moment on the particle. The particle reorientation depends on many factors like, the dielectric properties of both the particle and the host fluid, the geometry of the microparticle, the host viscosity, and the electric field.

Modification of PCLC microparticles by doping with conductive or dielectric dopants increases the particle conductivity and/or dielectric constant. Doping of PCLC particles reduces the reorientation time and extends the response to the DC regime operation. Doping with conductive dopants produced PCLC microparticles with a net charge, and they exhibited either translational or rotational motion depending on both the distribution of dopant within the particle and the dielectric characteristics of the host fluid. This behaviour observed by the author was novel and could find specific active applications.

The concept of electronic paper and sensors based on PCLC microparticles is very attractive to consumers nowadays. It offers possibility for a thin, reflective, lightweight, flexible devices that requires low energy consumption. In information technology as reflective multi-color information displays, flexible media for information display on either flat or curved surfaces. This PCLC technology could fit also to other applications, like switchable and tunable color filters, optical retarders, and micropolarizers, reflection and transmission properties for electro-optical and photonics applications; vehicle camouflage, patterned particles for storage of encoded and encrypted information for the security or military sector; switchable 'paints' or conformal coatings, 'smart windows' for energy or privacy control.

ACKNOWLEDGMENTS

This work was done at the University of Rochester, Laboratory for Laser Energetics in Rochester, NY, USA. The author thanks to Prof. Steve D. Jacobs for his guidance and supervision through this project, Mr. Kenneth Marshall and Dr. Tanya Kosc for their constructive advice and help. Author, also acknowledge the Laboratory for Laser Energetics for generous Horton Fellowship support.

REFERENCES

[1] Collings, P.; Hird, P. *Introduction to Liquid Crystals: Chemistry and Physics;* Taylor & Francis Ltd. London, UK, 1997.

[2] Yeh, P.; Gu, C. *Optics of Liquid Crystal Displays;* Wiley & Sons: NY, 1999.

[3] Chandrasekhar, S. *Liquid Crystals,* Cambridge University Press: NY, 1977.

[4] Korenic, E. M.; Jacobs, S. D.; Faris, S. M.; Li, L. *Mol. Crys. Liq. Crys.* 1998, *317*, 197-219.

[5] Trajkovska-Petkoska, A.; Jacobs, S. D.; Kosc, T. Z.; Marshall, K. L. *Manufacture of Regularly Shaped Polymer Cholesteric Liquid Crystal Flakes with a Mechanically Flexible Mold,* U.S. Pat. 7,238,316 B2, 2007.

[6] Trajkovska Petkoska, A. *Polymer Cholesteric Liquid Crystal Flakes – Their Electro Optic-Behaviour for Potential E-Paper Application*, VDM Verlag Dr. Müller, Saarbrücken,Germany, 2008.

[7] Trajkovska-Petkoska, A.; Varshneya, R.; Kosc, T. Z.; Marshall, K. L.; Jacobs, S. D. *Adv. Funct. Mater.* 2004, *15*, 217.

[8] Trajkovska-Petkoska, A.; Jacobs, S. D. *J. Mat. Sci. Eng. (A&B),* A2 (2), 2012, 137-151.

[9] Kosc, T. Z.; Marshall, K. L.; Trajkovska-Petkoska, A.; Kimball, E.; Jacobs, S. D. Displays 2004, *25*, 171–176.

[10] Trajkovska-Petkoska, A. *Polymer cholesteric liquid crystal flakes as new candidates for display and sensor applications.* NATO Science for Peace and Security Series-B: Physics and Biophysics: Nanotechnological Basis for Advanced Sensors, Springer, 2011.

[11] Wacker-Chemie, Consortium für Electrochemische Industrie GmbH, Zielstattstrasse 20, D-81379 München, Germany.

[12] Kosc, T. Z.; Marshall, K. L.; Jacobs, S. D.; Lambropoulos, J. C. *J. App. Phys.* 2005,*98*(1), 013509.

[13] Trajkovska-Petkoska, A.; Kosc, T. Z.; Marshall, K.; Hasman, K.; Jacobs, S.D. *J.Appl. Phys.* 2008, *103*, 094907.
[14] Trajkovska-Petkoska, A.; Jacobs, S. D.; Marshall, K. L.; Kosc, T. Z. *Electrically Actuated Doped Polymer Flakes and Electrically Addressable Optical Devices Using Suspensions of Doped Polymer Flakes in a Fluid Host* U.S. 7,713,436 B1, 2010.
[15] Maaroufi, A.; Haboubi, K.; El Amarti, A.; Carmona, F. *J. Mat. Sci.* 2004, *39*(1), 265-270.
[16] Grannan, D. M.; Gargland, J. C.; Tanner, D. B. *Phys. Rev. Lett.* 1981, *46*(5), 375-378.
[17] Zois, H.; Apekis, L.; Mamunya, Y. P. *Macrom. Sym.* 2003, *194,* 351-359.
[18] Carmona, F. *Physica A* 1989, *157*(1), 461-469.
[19] Moniruzzaman, M.; Winey, K. I. *Macromol.* 2006, *39*(16), 5194-5205.
[20] Tchmutin, I. A.; Ponomarenko, A. T.; Krinichnaya, E. P.; Kozub, G. I.; Efimov, O. N. *Carbon* 2003, *41*(7), 1391-1395.
[21] Schueler, R.; Petermann, J.; Schulte, K.; Wentzel, H. P. *J. App. Pol. Sci.* 1997, *63*(13), 1741-1746.
[22] Lux, F. *J. Mat. Sci.* 1993, *28*, 285-301.
[23] Martin, C. A.; Sandler, J. K. W.; Haffer, M. S. P.; Schwarz, M. K.; Bauhofer, W.; Schulte, K.; Windle, A. H. *Comp. Sci. Tech.* 2004, *64*(15), 2309-2316.
[24] Balberg, I. *Carbon* 2001, *40*, 139-143.
[25] Sandler, J. K. W.; Kirk, J. E.; Kinloch, I. A.; Shaffer, M. S. P.; Windle, A. H. *Polymer* 2003, *44*(19), 5893-5899.
[26] Jiang, X.; Bin, Y.; Matsuo, M. *Polymer* 2005, *46*, 7418-7424.
[27] Du, F. M.; Fischer, J. E.; Winey, K. I. *Phys. Rev. B* 2005, *72*(12), 121404.
[28] Miyasaka, K.; Watanabe, K.; Jojima, E.; Aida, H.; Sumita, M.; Ishikawa,K. *J. Mat. Sci.* 1982, *17*(6), 1610-1616.
[29] Flandin, L.; Prasse, T.; Schueler, R.; Schulte, K.; Bauhofer, W.; Cavaille, J. Y. *Phys. Rev. B* 1999, *59*(22), 14349-14355.
[30] Ounaies, Z.; Park, C.; K. E.Wise, K. E.; Siochi, E. J.; Harrison, J. S. *Comp. Sci. Tech.* 2003, *63*(11), 1637-1646.
[31] Barrau, S.; Demont, P.; Peigney, A.; Laurent, C.; Lacabanne, C. *Macromol.* 2003, *36*(14), 5187-5194.
[32] Kilbride, B. E.; Coleman, J. N.; Fraysse, J.; Fournet, P.; Cadek, M.; Drury, A.; Hutzler, S.; Roth, S.; Blau, W. J. *J. App. Phys.* 2002, *92*(7), 4024-4030.
[33] Balberg, I. *Phys. Rev. Lett.* 1987, *59*(12), 1305-1308.
[34] Balberg, I. *Carbon* 2002, *40*(2), 139-143.
[35] Breuer, O.; Sundararaj, U. *Pol. Comp.* 2004, *25*(6), 630-645.
[36] Dresselhaus, M. S.; Dresselhaus, G.; Jorio, A. *Ann. Rev. Mat. Res.* 2004, *34*, 247-278.
[37] Clayton, L. M.; Sikder, A. K.; Kumar, A.; Cinke, M.; Meyyappan, M.; Gerasimov, T. G.; Harmon, J. P. *Adv. Func. Mat.* 2005, *15*(1), 101-106.
[38] Watson, K. A..; Ghose, S.; Delozier, D. M.; Smith, J. G.; Connell, J. W. *Polymer* 2005, *46*(7), 2076-2085.
[39] Mitov, M.; Portet, C.; Bourgerette, C.; Snoeck, E.; Verelst, M. *Nature Materials* 2002, *1*(4), 229-231.
[40] Zois, H.; Apekis, L.; Mamunya, Y. P. *J. App. Pol. Sci.* 2003, *88*(13), 3013-3020.
[41] Ng, C. B.; Schadler, L. S.; Siegel, R. W. *Nanostructured Materials* 1999, *12*(1-4), 507-510.

[42] Mitov, M.; de Guerville, F.; Bourgerette, C. *Mol. Crys. Liq. Crys.* 2005, *435*, 673.

[43] Chiang, C. K.; Popielarz, R. *Ferroelectrics* 2001, *275*, 1-9.

[44] Cho, S. D.; Lee, S. Y.; Hyun, J. G.; Paik, K. W. *J. Mat. Sci.-Mat. Electronics* 2005, *16*(2), 77-84.

[45] Musal, H. M.; Hahn, H. T.; Bush, G. G. *J. App. Phys.* 1988, *63*(8), 3768-3770.

[46] Anantharaman, M. R.; Sindhu, S.; Jagatheesan, S.;. Malini, K. A.; Kurian, P. *J. Phys. D-App. Phys.* 1999, *32*(15), 1801-1810.

[47] Trajkovska-Petkoska, A.; Jacobs, S. D. *Mol. Cryst. Liq. Cryst.*, 2007.

[48] Marshall, K. L.; Hasman, K.; Leitch, M.; Cox, G.; Kosc, T. Z.; Trajkovska-Petkoska, A.; Jacobs, S. D. SID 07 Digest, editor J. Morreale, Society for Information Display, San Jose, CA, 2007, Vol. XXXVIII, Book II, 1741–1744.

In: New Developments in Liquid Crystals Research
Editors: A. T. Petkoska and A. T. Broach

ISBN: 978-1-63482-456-9

Chapter 3

MICROENCAPSULATION EFFECTS ON THE ELECTRO-OPTIC BEHAVIOR AND A REVERSE DRIVE MECHANISM FOR POLYMER CHOLESTERIC LIQUID CRYSTAL FLAKES

Gerald Philip Cox[*]
Corning Incorporated,
Senior Development Scientist, Surface & Coating Technologies
Corning Specialty Materials – Advanced Optics Division

ABSTRACT

A method for modeling the effect of microencapsulation on the electro-optical behavior of polymer cholesteric liquid crystal (PCLC) flakes suspended in a host fluid is introduced. Several microencapsulation configurations in an applied AC electric field are investigated using COMSOL MULTIPHYSICS software in combination with an analytical model. The field acting on the flakes is significantly altered as various encapsulant materials and boundary conditions are explored. The modeling predicts that test cells with multiple materials in the electric field path can have a wide range of electro-optic responses in ac electric fields. Both theoretical predictions and experimental evidence show that for PCLC flake reorientation to occur due to Maxwell-Wagner polarization, a reasonably strong electric field must be present along with at least moderately dissimilar PCLC flake and host fluid material dielectric constants and conductivities. For materials with low dielectric constants, electrophoretic behavior is observed under DC drive conditions at high field strengths for all evaluated microencapsulation configurations. This modeling method is shown to be a useful predictive tool for developing switchable particle devices that use microencapsulated dielectric particles in a host fluid medium. Also presented, is a means to orient reflective polymer cholesteric liquid crystal flakes both parallel and perpendicular to an electric field by doping of a low dielectric property host fluid. Doping of the siloxane host fluid with a small amount of water and salt creates a flake/host fluid suspension capable of

[*] Corresponding Author address: Email: CoxGP@Corning.com.

exhibiting a turnover frequency at low AC frequency and voltage. Above the turnover frequency the flakes orient perpendicular to the applied electric field and present a reflective surface to the incident light. Below the turnover frequency the flakes orient parallel to the electric field extinguishing the reflectance.

Keywords: PCLC flakes, polymer cholesteric liquid crystal flakes, reverse drive mechanism

1. Introduction

Polymer cholesteric liquid crystal PCLC flakes are micrometer-scale platelets of PCLC material either generated by freeze fracturing of well-aligned, environmentally robust macroscopic PCLC films or formed in controlled shapes and sizes by means of a number of photolithographic, molding, or stamping techniques (Fig. 1)[1-3]. Both processes rely on shear stress applied to the PCLC material surface during film or particle formation to align the helical structure normal to the film surface. The unique temperature stability and optical properties of the parent films are preserved. First developed in the 1990s for passive optical applications, PCLC flakes display a Bragg-like selective reflection effect, where incident light of a specific wavelength and circular polarization component is strongly reflected from the flake to produce highly saturated, circularly polarized colors [1]. Selective reflection occurs when the incident light satisfies the condition $\lambda = np$, where n and p are the average refractive index and the helical pitch length, respectively, of the PCLC material. This "selective reflection" is due to the helical structure inherent to the PCLC material and can be designed to reflect either left- or right-handed circularly polarized light, depending on the molecular structure of the PCLC polymers from which they are comprised. The individual color states depend on the flake's helical pitch length, which may be tuned from the deep UV nanometer-scale to the IR micrometer-scale region, including the entire visible spectrum. The polymeric structure effectively "locks in" the helical pitch, making the selective reflection colors insensitive to temperature over a very broad temperature range.

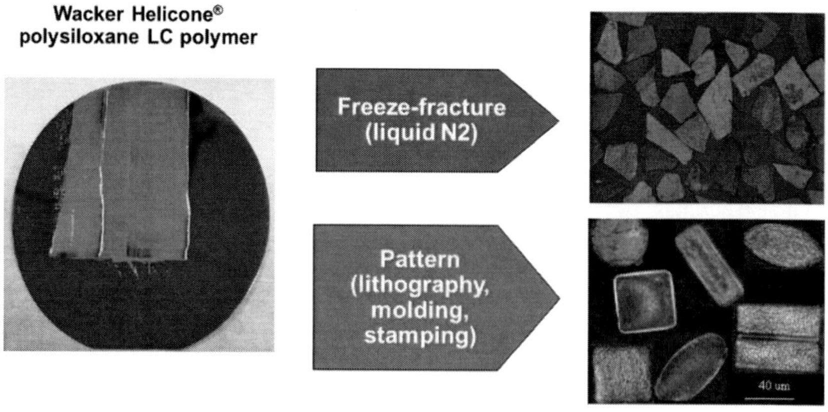

Figure 1. Methods for forming PCLC flakes either from well aligned parent films: blade-coated film on a 4-in.-diameter polished silicon wafer (left), freeze fractured into flakes (top right), or shaped flakes made by a soft lithography (bottom right).

1.1. Application of PCLC flakes to Switchable Particle Devices

Switchable particle-based technologies are of increasing interest for a number of applications as industry looks to develop products with unique optical properties and capabilities. An obvious application for reflective particles is information displays that require low power consumption and high brightness. As a result of their intrinsic selective reflection, PCLC flakes can provide bright, saturated colors without the need for additional color filters and polarizers [4]. Composite PCLC flake systems composed of either stacked left- and right-handed PCLC materials or a half-wave-plate material sandwiched between two materials of the same handedness can potentially lead to particle displays with reflectivities exceeding 80% [5]. A PCLC particle display does not require backlighting in daytime light conditions and would be uniquely suited for point-of-sale devices, portable devices, and flexible media on either flat or curved surfaces e.g., large-area signs, automobile dashboards, heads-up displays, and electronic paper. Other potential applications include switchable and tunable devices for color manipulation i.e., switchable or tunable color filters, switchable and tunable optical retarders or modulators for polarized light at desired wavelengths or bandwidth, switchable micropolarizers, switchable "smart windows" for either energy or privacy control, switchable coatings for use in decorative applications, and switchable coatings for applications in military security, camouflage, substrate reflectance control, document security, anticounterfeiting, and object tagging or identification [6].

1.2. Electro-Optical Behavior of PCLC flakes

Kosc et al. were the first to investigate the switchable behavior of irregularly shaped PCLC flakes in an applied electric field [7-10]. When flakes in an appropriate host fluid are subjected to an ac electric field, flake reorientation results due to Maxwell-Wagner (MW) interfacial polarization. Charges accumulate at the interface of the flake and the fluid, inducing a dipole on the flake. The applied electric field then acts on that dipole to reorient the flake parallel to the electric field. Interfacial polarization is driven by the difference between the flake and fluid dielectric constants and conductivities. Devices containing these "polarizing pigments" suspended in a commodity dielectric fluid host (e.g., silicone oil) at concentrations ranging from 3 to 5 wt % switch rapidly (<1 s) at very low voltages (10 to 100 $mV_{rms}/\mu m$) (Refs. 4 and 11). Figure 2 shows the electric field-induced behavior of a PCLC flake/fluid host suspension in a typical, sandwich-cell geometry [9,12]. Figure 2(a) indicates the dimensions of the rectangular-shaped, 3:1 aspect-ratio polysiloxane PCLC flakes used for the work reported here [2]. With no field applied, the flakes lie nearly parallel to the substrates and selectively reflect one circularly polarized component of the incident light [Fig. 2(b)]. With an electric field applied, patterned electrodes produce flake reorientation by MW polarization and extinguish the selective reflection color exposing the black back plane of the device [Fig. 2(c)]. The angle of rotation can range from 10° to 90°, depending on the dielectric constant and conductivity of both the flakes and the fluid host; 15° of rotation is often sufficient to extinguish the reflectivity.

Trajkovska-Petkoska *et al.* extended this work to shaped flakes, layered flakes, and doped flakes [3,13,14]. Uniformly shaped flakes lead to more-uniform reorientation times in an applied electric field. When the difference between host and flake conductivities is increased by doping, faster reorientation times due to both rotational and translational motions in a dc regime are observed. A combination of translational and rotational motions is observed when the dopant is not uniformly distributed throughout the particle. In moderately conductive hosts, 90° flake orientation is observed in the dc regime. Trajkovska-Petkoska *et al.* also expanded upon the original reorientation time model developed by Kosc *et al.* by including a gravity term for modeling flake relaxation times upon the removal of the electric field [13,15]. This expanded analytical model was used for the calculations presented here.

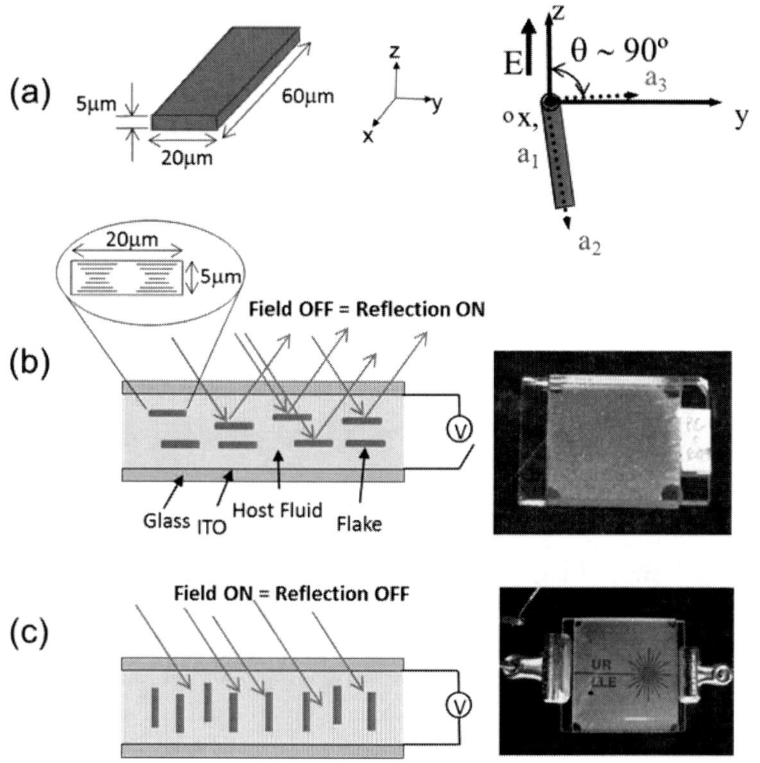

Figure 2. (a) Dimensions and coordinate system of a shaped PCLC flake. (b) Flakes lie approximately parallel to the cell substrates when no electric field is applied and appear colored as a result of selective reflection caused by the helical molecular structure of cholesteric liquid crystals, as depicted by the enlarged cross-sectional view of a flake. The longest axis is drawn into the page. (c) Flakes reorient with one long axis parallel to the applied field. As viewed from the incident light direction, rotated flakes appear dark since light is absorbed by the black back plane.

1.3. Microencapsulation of PCLC flake/fluid Host Suspensions

For PCLC flakes to achieve their potential in many switchable particle device applications, high-volume and low-cost manufacturing techniques such as roll-to-roll processing can be developed by microencapsulation [16-23]. Microencapsulation as applied to PCLC flakes involves suspending the PCLC flakes in a host fluid and then encapsulating this flake/fluid host suspension by either (1) forming discrete, optically transparent shells capsules from a dilute polymer encapsulation medium that can in turn be dispersed into the body of a film-forming transparent polymer or (2) emulsifying the suspension directly into the body of the film-forming polymer to form discrete capsules. Microencapsulation is necessary to prevent flake migration or agglomeration and to allow one to apply the flake/fluid host system to any surface rigid, flexible, or curved. Application may take place by a number of means, including knife coating, die coating, roll coating, and printing via ink-jet techniques.

A method is presented here to model and predict the electro-optic EO behavior of microencapsulated, shaped PCLC flakes. Experimental results are presented to confirm the validity of the modeling as a tool for developing systems for switchable-device applications.

1.4. Reverse Drive Mechanism for PCLC flake/fluid Host Suspensions

As shown in Fig. 3, in order for the flakes to orient perpendicular to the electric field, gravity must be counted on to relax the flake back into a reflective position. Relaxation times vary (6 to >60 s) depending on the density differences between the flake and host fluid as well as the viscosity of the host fluid. Recent work has focused on alternate host fluids with different dielectric properties, as well as the effects of changing the dielectric properties of the PCLC flakes themselves through doping [13, 24]. The different system combinations compatible with MW reorientation have used various factors to adjust the speed of PCLC flake reorientation parallel to the electric field, but have failed to control relaxation outside of the use of gravity. If the display itself is not oriented perpendicular to gravity, the flakes may not relax to a full reflective position. Therefore, in order for PCLC flake technology to reach its full potential in a reflective display application, it is necessary that a reverse drive mechanism be developed, independent of gravity, to reorient the PCLC flakes to an orientation perpendicular to the electric field.

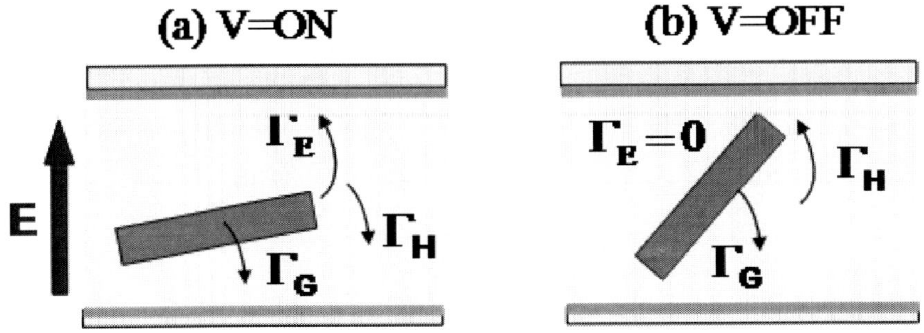

Figure 3. Illustration of torques acting on a PCLC flake in an electro-optic test cell when the flake density is greater than that of fluid. (a) When an electric field is applied an electrostatic torque acts to orient the flake in line with the electric field and the torques due to gravity and hydrodynamic drag resist that orientation. (b) When the electric field is turned off, the torque due to gravity acts to reorient the flake and hydrodynamic drag resists that motion. Adapted from reference 25.

2. Modeling and Devices

2.1. Materials Systems

The materials presented in this chapter were chosen for a number of compelling reasons. All the materials selected are available commercially, in large quantities, at low cost, and are essentially nontoxic. The host fluids were chosen to span a wide range of dielectric constants and conductivities to include a range of small to large electric field interactions. A subset of these fluids is further modified to address the need for a reverse drive mechanism. The studied host fluids were doped for increased fluid conductivity while maintaining a low dielectric constant. Both the aqueous and nonaqueous polymer binder solutions employed for microencapsulation are easily emulsified and cast with standard laboratory equipment. The photoresist and UV curable epoxy were selected because of their resistance to the host fluids chosen and their ability to be processed with standard laboratory equipment. The materials presented in this chapter and their properties are listed in Tables I and II.

Table 1. Properties of materials used for modeling and/or experiments

Material	ε	σ (S/m)	ρ (kg/m^3)	η_0 (mPa·s)	Function
PCLC flakes[a]	2.2	1x10−9	1100	n/m[b]	Particle
SIT7757	2.7	3x10−8	1070	35	Host fluid
DMS-E09	7.0	9x10−8	995	9	Host fluid
Propylene carbonate	60.0	1x10−4	1200	2	Host fluid
SU-8 3050 photoresist	3.2	2.13x10−8	n/m	n/a[c]	Well/cube walls
SU-8 2015 photoresist	3.2	2.13x10−8	n/m	n/a	Cube top/bot
OG142-13 UV epoxy	6.9	2.8x10−7	n/m	n/a	Adhesive
PVA	2.0	1x10−9	n/m	n/a	Binder
Porcine gelatin	5.4	1.79x10−6	n/m	n/a	Capsule shell
Gum arabic	6.5	1.08x10−4	n/m	n/a	Capsule shell
PDMS	2.6	1x10−9	1030	n/a	Binder
(AOT) Aerrosol OT-100 surfactant	n/a	n/a	n/a	n/a	Dopant
NaCl	n/a	n/a	2165	n/a	Dopant

[a]Flake—L x H x D = 20 x 5 x 60 μm^3.

[b]n/m = not measured.

[c]n/a = not applicable.

Table 2. Properties of doped host fluids for modeling and/or experiments

Material %	AOT	% NaCl	% water	(S/m)
SIT7757/AOT/water	1.0	0	0.5	3.18×10^{-8}
SIT7757/NaCl/water	1.0	0.025	0.475	8.00×10^{-6}

2.2. Modeling Methodology

The electric field that acts on the PCLC flake directly influences its EO behavior. In an ac electric field, the strength of the rotational torque applied to the flake depends on the following:

1) the effective electric field strength acting on the flake, and
2) the strength of the MW polarization induced on the flake surface, which is controlled by the following:
   a. the difference between the dielectric constants of the flake and host fluid, and
   b. the difference between the conductivities of the flake and host fluid.

When a dc field is applied, the EO behavior results from either interfacial polarization or electrophoresis. The dominating effect will depend on the following:

1) the intrinsic charge on the flake in the host fluid,
2) the strength of interfacial polarization induced on the flake surface by the difference between the dielectric constants of the flake and host fluid, and
3) the effective strength of the electric field acting on the flake.

The rotational reorientation of a PCLC flake is the result of a balance of torques applied to the flake [13,15]. The electro static torque Γ_E, from the applied electric field, acts to reorient the flake as shown when moving from the "field off" state [Fig. 2(b)] to the "field on" state [Fig. 2(c)]. The resistance to rotation is the sum of the torques caused by the hydrodynamic force (Γ_H) and gravity (Γ_G):

$$\Gamma_E=\Gamma_H+\Gamma_G \tag{1}$$

The Clausius-Mosotti factor (K_i^*) is a measure of the strength of the effective polarization of the flake. This factor incorporates the dielectric constant and conductivity of both the PCLC flake (ε_p and ε_p) and host fluid (ε_h and ε_h), the depolarization factor of the flake (A_i), as well as the frequency of the applied electric field (ω).

$$K_i^* = \frac{\left(\varepsilon_p-i\frac{\sigma_p}{\omega}\right)-\left(\varepsilon_h-i\frac{\sigma_h}{\omega}\right)}{\left(\varepsilon_h-i\frac{\sigma_h}{\omega}\right)+A_i\left[\left(\varepsilon_p-i\frac{\sigma_p}{\omega}\right)\left(\varepsilon_h-i\frac{\sigma_h}{\omega}\right)\right]} \tag{2}$$

The electrostatic torque that drives flake reorientation may be written as a function of the flake's semiaxes (a_i), the dielectric constant of the host fluid (ε_h), the imaginary portion of the Clausius-Mosotti factor (K_i^*), the depolarization factor of the flake (A_i), the strength of the electric field applied to the flake (E_0) , and the angle of the flake (θ) relative to the electric field.

$$\Gamma_{Ei} = \frac{4}{3}\pi a_i a_j a_k \varepsilon_h K_j^* K_k^* (A_k - A_j) E_0^2 \sin\theta\cos\theta$$
$$= C_{Ei}\sin\theta\cos\theta \tag{3}$$

The hydrodynamic torque (Γ_{Hi}) counteracts the electro static torque and may be defined as:

$$\Gamma_{Hi} = \frac{16}{3}\pi a_i a_j \eta_0 \frac{(a_j^2 a_k^2)}{(a_j^2 A_j + a_k^2 A_k)} \Omega_i = C_{Hi}\Omega_i \tag{4}$$

where a_i are the length of the flake semiaxes, η_0 is the host fluid viscosity, A_i is the depolarization factor of the flake, and Ω_i is the angular velocity of the flake about its axis. The torque caused by gravity (Γ_{Gi}) is specific to the configuration in which the experiments are conducted. For the work described in this chapter, the test devices lie flat and the force of gravity acts in the -*z* direction [Fig. 2(a)]. The gravitational torque is a function of the PCLC flake volume (V_p), density of both the flake (ρ_p)and host fluid (ρ_h), gravity (g), flake dimension (a_j), and the angle of the flake (θ) relative to the electric field [Fig. 2(a)].

$$\Gamma_{Gi} = V_p(\rho_p - \rho_h) g a_j \cos\theta \tag{5}$$

Substituting Eqs. (3)–(5) into Eq. (1), we can solve for the angular velocity of the flake

$$\Omega_i = \frac{d\theta}{dt} = \frac{C_{Ei}\sin\theta\cos\theta - C_{Gi}\cos\theta}{C_{Hi}} \tag{6}$$

Equation (6) may then be integrated for the time to reorient the PCLC flake from an initial angle θ_0 ~0° to its final position θ_f ~90° as:

$$\int dt = C_H \int \frac{d\theta}{C_E \sin\theta\cos\theta - C_G\cos\theta} \tag{7}$$

$$t = \frac{C_{Hi}}{C_{Ei}^2 - C_{Gi}^2}\left[(C_{Gi} - C_{Ei}) ln\left[\cos\frac{\theta}{2} + \sin\frac{\theta}{2}\right] - (C_{Ei} - C_{Gi}) ln\left[\cos\frac{\theta}{2} - \sin\frac{\theta}{2}\right] + C_{Ei} ln[C_{Gi} - C_{Ei}\sin\theta]\right]_{\theta_0}^{\theta_f} \tag{8}$$

To develop the analytical model shown above, a *basic test cell* with only a flake/fluid host suspension in the cell gap between the electrodes was studied in previous work to characterize PCLC flake motion in ac and dc electric fields (Fig. 2) [7,12-15]. This type of cell is presented in previous chapters and used as a baseline for the work reported here.

As the complexity of the cell design increases, the boundary conditions required for an analytical solution make it much more challenging to calculate the electric field applied to the flake. To account for the added boundary conditions created by the microencapsulation of the PCLC flake/fluid host suspension, another method must be found to model the electric field within the test cell. Finite element analysis-based software offers a straightforward method to account for these boundary conditions in the calculation of the electric field acting on the PCLC flake. The calculation of the electric field combined with the analytical model enables one to predict PCLC flake motion caused by MW polarization in an electric field for complex cell geometries.

Finite element analysis-based software, COMSOL MULTIPHYSICS (Comsol, Inc.), was used to analyze the electric field for six types of test cell configurations with three types of host fluids for each configurational variant. The six configurations studied are illustrated in Fig. 4. The *basic* cell [Fig. 4(a)] with only a flake/fluid host suspension in the cell gap is used as a baseline for the work presented here. The *microwell* cell [Fig. 4(b)] adds vertical walls composed of photoresist to the cell gap to constrain the flake/fluid host suspension. The fluid is still in contact with the electrodes as in the basic cell and the electric field experiences a *continuous* material path between the electrodes just as in the basic cell. The *microcube* cell [Fig. 4(c)] builds on the microwell configuration of Fig. 4(b) by adding a layer of photoresist above and below the host fluid that creates a *discontinuous* path for the electric field between the electrodes. For *direct encapsulation* cells, two possible methods of assembly are shown. Direct encapsulation type (I) [Fig. 4(d)] has the flake/fluid host constrained in a capsule. The path for the electric field is both discontinuous and nonuniform because the curved edges of the capsule are parts of the path between the electrodes. This configuration also includes a layer of epoxy adhesive used for assembly between the binder layer and the top electrode.

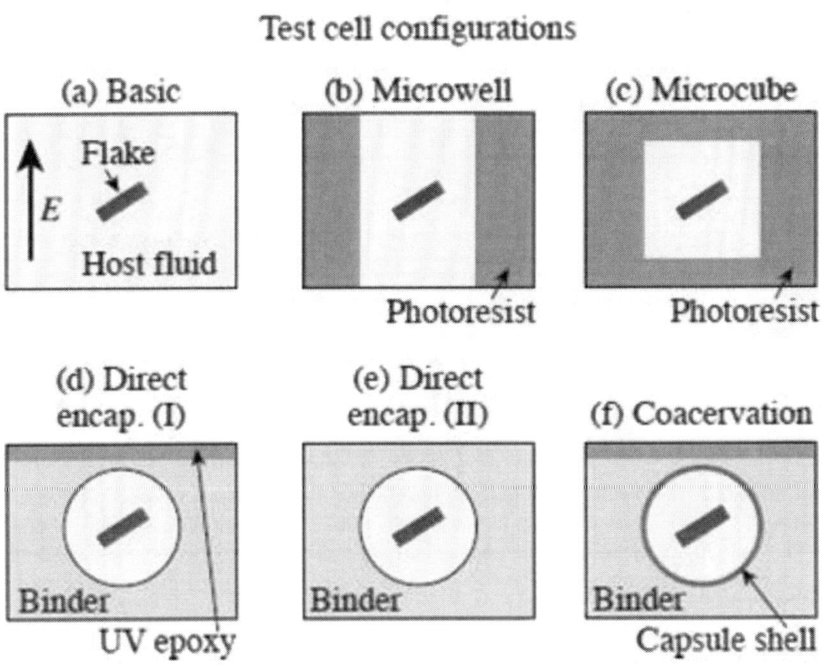

Figure 4. Six configurations of PCLC flake test cells studied. (a) Basic cell: Contains only a flake/fluid host suspension in the cell gap. (b) Microwell: Insulative vertical walls (photoresist) confine the flake/fluid host suspension within the well; fluid in contact with electrodes. (c) Microcube: The cube (photoresist) fully encapsulates the flake/fluid host suspension; fluid is not in contact with electrodes. (d) Direct encapsulation (I): A polymer binder (capsule) fully encapsulates the flake/fluid host suspension. The fluid is not in contact with the electrodes and there is an epoxy layer between the polymer binder and the top electrode. (e) Direct encapsulation (II): The polymer binder (capsule) fully encapsulates the flake/fluid host suspension. The fluid is not in contact with electrodes and there is no epoxy layer between polymer binder and top electrode. (f) Coacervation encapsulation: A polymer binder surrounds a capsule, a thin shell of material different than the binder layer, which fully encapsulates the flake/fluid host suspension. The fluid is not in contact with electrodes and there is an epoxy layer between the polymer binder and the top electrode.

Direct encapsulation type (II) [Fig. 4(e)] also has the flake/fluid host suspension constrained in a capsule similar to type (I) except there is no adhesive layer between the binder layer and top electrode. Finally, the *coacervation* cell [Fig. 4(f)] represents a device that, once assembled, is very similar to the direct encapsulation devices. Here there is an additional discontinuous material forming a separate capsule shell that further complicates the path of the electric field.

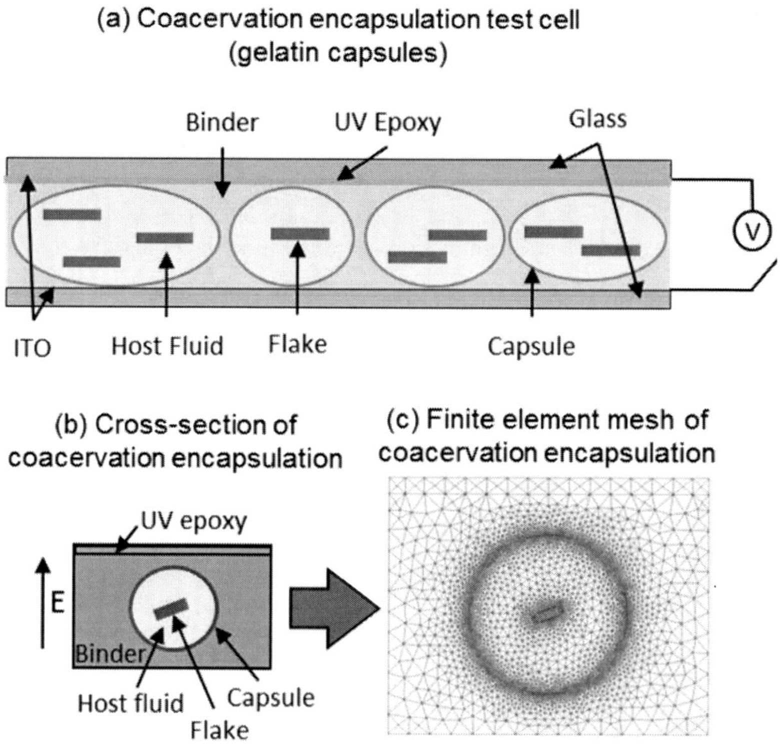

Figure 5. (a) Illustrated cross section of a gelatin-gum arabic microcapsule test cell, a more complex version of the configuration shown in Fig. 2 that is based on Fig. 3 (f). (b) The equivalent cross section for analysis with fluid, flake, gelatin-gum arabic capsule, UV epoxy, and polymer binder boundaries in the cell gap. (c) Finite-element mesh of the coacervation test cell cross section generated with COMSOL MULTIPHYSICS.

The different encapsulation types are intended to show an evolution of complexity in moving from a test cell with only a flake/fluid host suspension in an applied electric field to a flake/fluid host suspension microencapsulated within discrete gelatin capsules. Figure 5(a) illustrates a coacervation-type test cell (gelatin capsules) used for EO characterization; the gelatin capsules have been dispersed into a separate film-forming polymer binder, coated onto an indium tin oxide (ITO)-coated substrate and then attached to a second substrate with an UV-cured adhesive. Figure 5(b) represents the two-dimensional (2D) cross section of the cell to be modeled. Figure 5(c) is the matching finite-element mesh generated within COMSOL MULTIPHYSICS, using the "in-plane electric currents" module to analyze the electric field. The gelatin-type capsule with an UV adhesive layer in this configuration is the most complex material set studied described in this chapter. To model each of the test configurations, a representative 2D cross section is generated using the COMSOL MULTIPHYSICS module. Each subdomain (geometric section) is assigned a value for conductivity, dielectric constant, and thickness in the $-x$ direction (into the plane of the page, as shown in Fig. 2). The glass substrates and ITO layers are not included in the model. An ac voltage (7.5 V_{rms} unless noted) is applied to the lower boundary (electrode) of the model and the top boundary (electrode) is set to ground potential. The side boundaries are defined as electrically insulating. The ac frequency is set at 100 Hz and the overall cell gap is 150 μm unless noted otherwise. The flake angle, with respect to the electrode, is fixed at 20° to calculate the electric field for all configurations. The critical cell dimensions, materials, and their properties used in each configuration are listed in Tables 1, 2 and 3. The flake dimensions are as shown in Fig. 2(a). For direct comparison of microencapsulation effects on the electric field, only the materials properties of SU-8 3050 photoresist were used to model photoresist as well as binder layers, even though experiments were conducted with binders made from SU-8 3050, SU-8 2015, polyvinyl alcohol (PVA), and polydimethylsiloxane (PDMS). All four materials have low conductivity and dielectric constant; using only SU-8 3050 for modeling makes the comparison between microencapsulation types straightforward and does not significantly affect the results. As shown in Fig. 5 (c), the model is divided into a continuous mesh of finite elements. Each configuration includes a subdomain of the host fluid in a 1-μm-thick area immediately surrounding the flake. The normal component of the electric field {e.g., E_0 = sqrt[$(E_x)^2 + (E_y)^2$]} is integrated over this subdomain to calculate the value of the electric field acting on the flake due to the boundary conditions for each configuration. This method is a reasonable approximation for E_0 used in the analytical model because $E_z \sim 10E_y$. For the basic cell, this integration is not necessary because the medium in the cell gap (and thus the electric field) surrounding the flake is homogenous, but this step becomes important when the flakes are encapsulated. Once the electric field acting on the flake has been calculated, this value is inputted into the analytical model developed previously [13,15]. The analytical model results are then calculated using MATHEMATICA software Wolfram Research, Inc. to evaluate the PCLC flake reorientation times as a function of the electric field's amplitude and frequency and the physical properties of the flake/fluid host system [13,15].

2.3. Device Preparation

A minimum of two prototype test cells for each of the six configurations shown in Fig. 3 were prepared and evaluated for EO flake behavior. Each prototype was similar to the devices shown in Fig. 2, but without patterned electrodes. All of the devices were made using rectangular, 3:1 aspect-ratio polysiloxane PCLC flakes [see Fig. 2(a)] [2]. Identification of sources for all materials and preparation methods for each configuration are given below.

The PCLC material is a non-cross-linked cyclic polysiloxane (Wacker-Chemie GmbH, Munich, Germany) with a glass transition temperature T_g between 40 and 50 °C. The PCLC flakes were made by soft lithography using a patterned silicon wafer (Semiconductor/ Microsystems Fabrication Laboratory, Rochester Institute of Technology, Rochester, NY) to make a mold from a two-part PDMS molding formulation (Dow Corning Corp., Sylgard, Midland, MI). The PCLC material was blade coated into the mold at 195 °C (Refs. 2 and 3).

Table 3. Nominal dimensions and materials used for modeling and manufacturing of the six experimental test cell configurations. (As-built dimensions are similar)

							Epoxy Capsule Shell		layer	Top/bottom	
Configuration	Cell gap (μm)	Wall Height (μm)	Wall width (μm)	Top/bottom thick (μm)	Diam (μm)	Capsule Thick (μm)	thick (μm)	Wall material	layer material	Binder material	Capsule material
Basic[a]	150	n/a[b]	n/a	n/a	n/a	n/a	n/a	n/a	n/a	n/a	n/a
Microwell[a]	150	150	50	n/a	n/a	n/a	n/a	SU-8	n/a	n/a	n/a
Microcube[a]	150	150	50	20	n/a	n/a	n/a	SU-8	SU-8	n/a	n/a
Direct Encapsulation I [c,d]	190	n/a	n/a	n/a	100	n/a	40	n/a	n/a	PVA/SU-8	n/a
Direct Encapsulation II [e,d]	150	n/a	n/a	n/a	100	n/a	n/a	n/a	n/a	PDMS/SU-8	n/a
Coacervation[c]	190	n/a	n/a	n/a	100	1	40	n/a	n/a	PVA	Gelatin-gum arabic

[a]SIT7757, DMS-E09, PC host fluids tested.
[b]n/a=not applicable.
[c]SIT7757, DMS-E09 host fluids tested.
[d]SU-8 3050 material properties used for modeling to allow one to directly compare microencapsulation types.
[e]PC host fluid tested.

Propylene carbonate (PC) (99.5%, Acros Organics, Fair lawn, NJ) was used as a fluid host with high dielectric constant and conductivity. DMS-E09 (Geleste, Inc., Morrisville, PA) was used as a fluid host with moderate dielectric constant and conductivity. SIT7757 (Geleste, Inc., Morrisville, PA) was used as a host fluid with low dielectric constant and conductivity.

SIT7757 was also used as a doped host fluid with increased conductivity. Aerosol OT-100 (Cytec Industries, Woodland Park, NJ) was a surfactant used as a dopant to increase the conductivity of the host fluid. NaCl (Aldrich Chemical Co., Inc., Milwaukee, WI) was also used as a dopant to increase the conductivity of the host fluid.

Electro-optical test cells were fabricated using 25 x 25 x 3 mm ITO-coated substrates (Thin Film Devices, Inc., Anaheim, CA), cleaned by a detergent wash (Extran MA 02, EMD Chemicals Inc., Gibbstown, NJ). The basic test cells used 80 μm microspheres (Duke Scientific Corp., Palo Alto, CA) mixed into EPO-TEK OG 154 UV curing epoxy (Epoxy Technology, Billerica, MA) to set the test cell gap spacing. A drop of OG-154 epoxy/microspheres mixture was applied to the corners of the cell, with the substrates offset so that clips could be attached for power. The ac power was supplied from a Stanford research function generator, model DS345 (Sunnyvale, CA), and an HP 6827A bipolar power supply/ amplifier (Hewlett-Packard Co., Palo Alto, CA). The dc power was supplied by a Fluke high voltage power supply, model 412B (Fluke Corp., Everett, WA). A Blak Ray Long- wave UV lamp, model B 100AP (San Gabriel, CA) was used to cure the epoxy at 365 nm. The two outside edges of the cell were sealed with epoxy and cured. The cell was filled with flake/fluid host suspension (1% flakes) by capillary action and the remaining two edges sealed with epoxy and cured.

Microwell cells were made by spin coating SU-8 3050 photoresist (MicroChem, Newton, MA) at 1000 rpm onto ITO-coated substrates. The microwells were photopatterned into the SU-8 3050 using a chrome mask (Semiconductor/ Microsystems Fabrication Laboratory, Rochester Institute of Technology, Rochester, NY) and Blak Ray Longwave UV lamp, model B 100AP (San Gabriel, CA). The wells were developed and edge bead was removed using MicroChem Developer Solution (MicroChem, Newton, MA). Next, a drop of OG-154 epoxy without microspheres was applied to the corners of the cell. The microwells were filled with a flake/fluid host suspension by flood coating the microwells and using the top substrate to squeeze out the excess flake/ fluid host suspension to an unpatterned area. The cell gap is set by the height of the microwells. The substrates were aligned such that the top and bottom substrates were offset so that clips could be attached for power from the Stanford Research function generator and ac amplifier or dc power supply. The OG 154 epoxy at the corners of the cell was partially cured; then the cell was edge sealed with epoxy and fully cured.

The microcube cells are constructed following the same procedure as the microwell cells except that a 20 μm layer of SU-8 2015 photoresist (MicroChem, Newton, MA) was spin coated onto the top and bottom substrates, exposed, and developed prior to photopatterning the microwells onto the bottom substrate. The assembly procedure for the microcube devices is the same as that for the microwell devices described earlier.

Direct encapsulation cells were made by emulsifying the flake/fluid host suspension (1% PCLC flakes) into a 20% aqueous solution of 80% hydrolyzed PVA (Aldrich Chemical Co., Inc., Milwaukee, WI). Emulsification was accomplished with low shear by adding 1 g total of the components into a 4 ml vial and rotating at 45 rpm for 15 min. A portion of the emulsion, which depends on the size of the substrate to be coated, was then cast by hand onto an ITO-coated substrate and left to dry for 24 h. Once dry, OG142-13 UV epoxy (Epoxy Technology, Billerica, MA) is applied to the top of the PVA film; the second substrate is then applied and the cell cured with UV light (365 nm).

The complex coacervation cells were made in two steps: (a) the flake/fluid host suspension was microencapsulated by complex coacervation into gelatin-gum arabic capsules; (b) the capsules were dispersed into a polymeric binder and made into a test cell. Microencapsulation is accomplished by emulsifying 3 g of the flake/fluid host suspension into 6 g of a 1% aqueous solution of gum arabic (Aldrich Chemical Co., Inc., Milwaukee, WI) at 55 °C using a high-shear mixer (IKA Ultra Turrax Tube Disperser, Cole-Parmer Instrument Co., Vernon Hills, IL). Then 6 g of a 1% solution of porcine gelatin (Aldrich Chemical Co., Inc., Milwaukee, WI) was added, further emulsified, and temperature reduced to 41 °C. Next, the emulsion was diluted by dropwise addition of 50 g of water. The *p*H was then reduced to 4 by dropwise addition of 30 g of a 0.2% aqueous solution of acetic acid to induce coacervation. Coacervation continued for 30 min and then the batch was cooled to 5 °C. Agitation was continued for 40 min and then the hardening agent, 10 g of a 5% aqueous solution of glutaraldehyde, was added dropwise. Agitation continued for 1 h and then 100 g of a 4% aqueous solution of NaOH was added dropwise to raise the *p*H to 10 and complete the hardening reaction. The batch of finished capsules was then warmed to room temperature, being dispersed with low shear into 1 g of a 20% aqueous solution of PVA. A film of the dispersion was then cast by hand onto an ITO substrate and the cell was prepared following the same procedure as a direct encapsulation cell.

The doped host fluids were prepared by emulsifying the SIT7757 siloxane fluids with formulations listed in Table 2. The dopants were added to the host fluid and mixed by manual shaking. The basic test cell design was used to characterize doped host fluids. The cells were filled with flake/doped fluid host suspension (1% flakes) by capillary action and the remaining two edges sealed with epoxy and cured.

2.4. Characterization

Each device was tested within 48 h of assembly to reduce the effect of aging on the test cell. Aging may increase the reorientation times of the PCLC flakes if the ITO electrodes are not insulated from the host fluid.[13] The flake motion and response times were characterized by streaming video with time stamping using a Leitz Orthoplan polarizing optical microscope (POM) (Wetzlar, Germany) in reflection mode coupled with a MicroPix C1024 (CCD) camera and Streampix software (CCD Direct, Ann Arbor, MI).

Optical images were taken using a Leica DMXR POM (Leica Microsystems, Wetzlar, Germany) in reflection mode coupled with a Q-Imaging Retiga 4000R Fast camera and QCAPTURE PRO software (Media Cybermetics, Inc., Bethesda, MD).

The dielectric properties of each test cell and materials were measured using a Solartron 1260 impedance gain-phase analyzer (Solartron Analytical, Farnborough, Hampshire, GU14 0NR, England) and collected with ZPLOT and ZVIEW (Scribner Associates, Inc., Southern Pines, NC) PC interface softwares.

3. RESULTS AND DISCUSSION

3.1. Effect of Cell Geometry on the Electric Field

The environment (cell geometry) surrounding the flake has a large effect on flake motion because it can influence the applied electric field strength, which is directly proportional to the electrostatic torque applied to the flake. We calculated the effect of the cell geometry (device configuration) on the electric field while keeping the properties of both the fluid and the flake constant.

Figure 6 shows the COMSOL MULTIPHYSICS modeling results for four configurations where a PCLC flake is suspended in the high dielectric constant host fluid PC in a 150-μm-thick test cell with an applied electric potential of 7.5 V. The material properties of the photoresist SU-8 3050 were used to model the microwell/microcube walls and the direct encapsulation binder to enable one to directly compare the electric fields between the device geometries. The resulting plots of the electric fields are shown as the test cell geometry changes from a basic configuration [Fig. 6(a)] to a direct encapsulation type-II cell [Fig. 6(d)].

When the path of the electric field between the electrodes encounters constant material properties, as in both the basic and microwell configurations, a strong electric field is present in the cell gap [Figs. 6(a) and 6(b)]. This is characterized by the light gray areas throughout most of the cell gap. A slight drop of the electric field in localized areas the flake is denoted by a slightly darker gray.

Additionally, if the dielectric constants and conductivities of the host fluid and PCLC flake are very different, the model predicts a high electrostatic torque on the flakes. When the path of the electric field between the electrodes does not encounter constant material properties, as in both the *microcube*- and direct encapsulation type-II configurations [Figs. 6(c) and 6(d)], the model prediction depends on the material properties of the system. When the electric field passes from a low dielectric constant material such as SU-8 3050 ($\varepsilon = 3.2$) into a high dielectric constant material such as PC ($\varepsilon = 60$), the electric field drops off dramatically.

This is seen as regions of very dark gray surrounding the flake in the microcube and direct encapsulation type-II examples [Figs. 6(c) and 6(d)]. In such cases, even when the dielectric constant and conductivity mismatch between PC and PCLC is large, the flake will experience low electric field strength and therefore low electrostatic torque. This represents a nonfunctional device configuration.

In addition, a distortion in the electric field results as it encounters material composition changes that are non-normal to the direction of the electric field propagation. This phenomenon is apparent in both the microcube example at the corners of the microcube and the direct encapsulation example near the edges of the capsule wall [Figs. 6(c) and 6(d)] , where the arrows denoting the electric field are at an angle to, rather than perpendicular to, the top electrode.

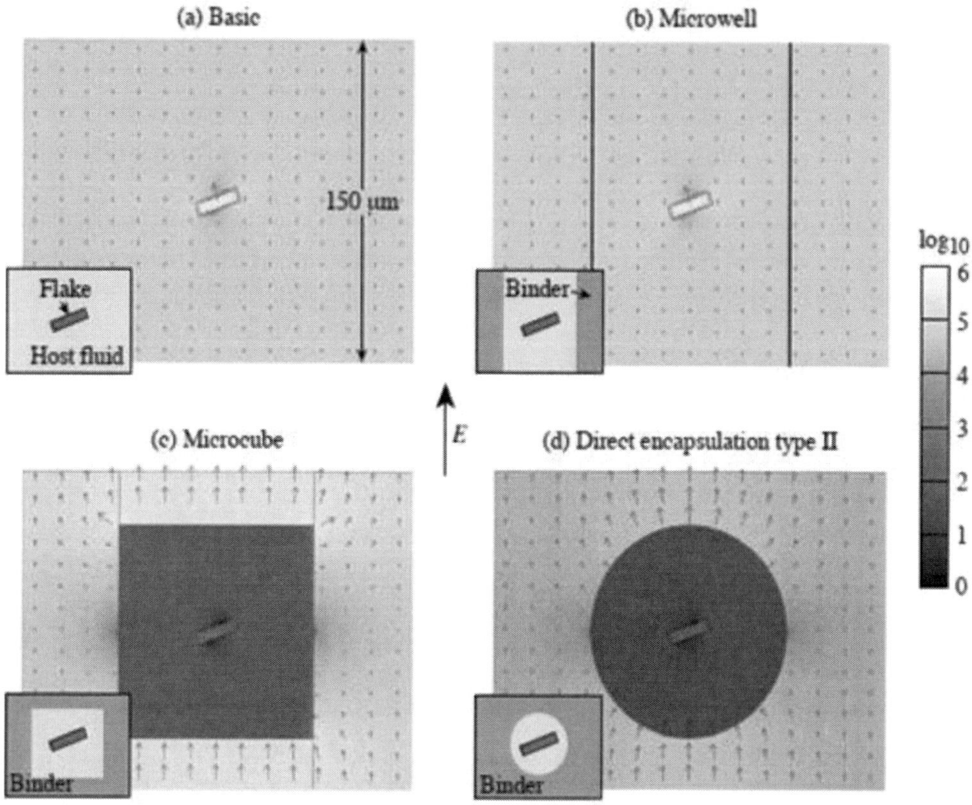

Figure 6. COMSOL MULTIPHYSICS output of the electric field as a function of test cell geometry (insets highlight geometry). The fluid surrounding the flake has the properties of PC ($\varepsilon = 60$), the flake is PCLC ($\varepsilon = 2.2$), the encapsulating material binder is SU-8 3050 ($\varepsilon = 3.2$), and the applied voltage to the 150 μm-thick cell is 7.5 V_{rms}. The direction and size of the arrows represent the direction and strength of the electric field, the surface plot density also represents the strength of the electric field, and the scale range is 0 (black) to 10^6 (white) V/m. (a) Basic and (b) microwell configurations have strong uniform electric fields because the materials in the path of the electric fields have uniform dielectric properties. (c) Microcube and (d) direct encapsulation configurations have nonuniform electric fields because materials in the path of the electric fields have dissimilar dielectric properties. Note the electric field distortion (arrow directions at an angle) near the corners and edges of the microcube and capsule in direct encapsulation.

3.2. Effect of Host fluid on the Electric Field

If the cell geometry is held constant and several host fluids are considered, an equally dramatic effect is observed in the effective electric field strength as it passes from region of low to high dielectric constant.

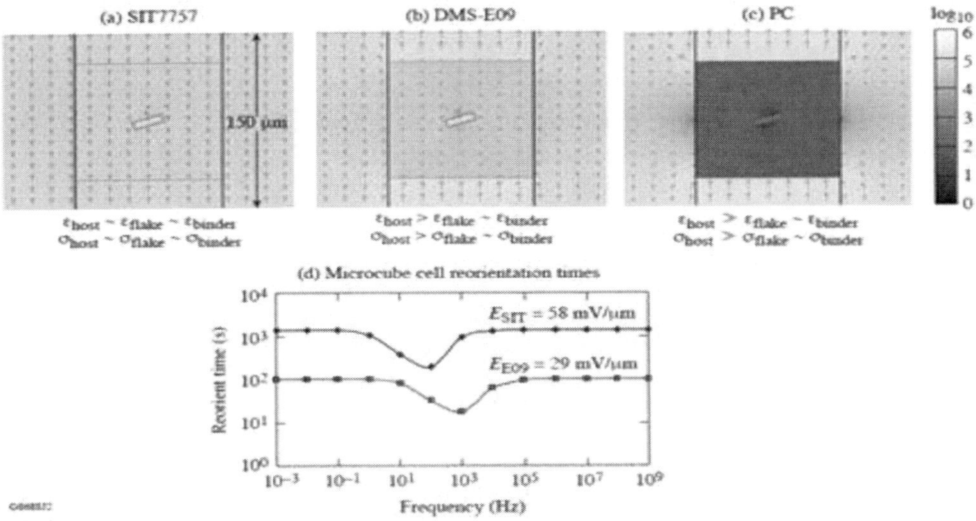

Figure 7. Electric field as a function of fluid properties in a microcube cell. The direction and size of the arrows represent the direction and strength of the electric field, the surface plot density also represents the strength of the electric field, and the scale range is 0 (black) to 10^6 (white) V/m. (a) Low ε, σ (SIT7757), (b) moderate ε, σ (DMS-E09), and (c) high ε, σ (PC). (d) A corresponding plot of predicted reorientation times for SIT7757 and DMS-E09 with an applied potential of 7.5 V_{rms} to a 150-μm-thick cell as a function of ac frequency and host fluid type. The terms E_{SIT} and E_{E09} are the calculated electric fields acting on the PCLC flake for each host fluid. No curve for PC is shown because E_{PC} is insufficient to reorient the flake.

In Fig. 7, from left to right, a microcube cell is shown with the fluid host properties changing from low ε and σ [e.g., SIT7757 $\varepsilon = 2.7$, $\sigma = 10^{-8}$, Fig. 6(a)] to moderate ε and low σ [e.g., DMS-E09 $\varepsilon = 7$, $\sigma = 10^{-8}$, Fig. 7(b)] to high ε and σ [e.g., PC $\varepsilon = 60$, $\sigma = 10^{-4}$, Fig. 7(c)]. For all three cells the applied electric potential is 7.5 V_{rms}, the cell thickness is 150 μm, and the flake angle was set to 20°. Field strength and, therefore, the color within the microcube and adjacent to the flake change from high (light gray) to moderately high (gray) to low (dark gray).

The effect on the magnitude of the electric field combined with the dielectric constant mismatch can be seen in Fig. 7 (d), where the predicted reorientation times are plotted corresponding to the microcube cells shown in Figs. 7(a) and 7(b). As seen in Fig. 7(d), SIT7757 has a higher electric field acting on the flake but with a similar dielectric constant to the flake a longer reorientation time results. DMS-E09, although it has a lower effective electric field, has a greater difference in dielectric constant with the flake and therefore a faster reorientation time. There is no curve plotted for PC as the host fluid in Fig. 7(c) because the electric field acting on the flake is so small that no reorientation takes place. This is a nonfunctional material combination.

The ac electric field has a range of frequencies over which the induced charge buildup at the flake/fluid host interface caused by MW polarization has the mobility needed to form an induced dipole on the flake. The induced is then acted on by the electric field to reorient the flake. In Fig. 7(d), when the frequency of the electric field is matched to the charge mobility, a minimum reorientation time is achieved. The particular range of frequencies at which reorientation occurs depends on the composition of the materials system, as seen in Fig. 7(d).

The SIT 7757 oil has a minimum predicted reorientation frequency of ~200 Hz, while DMS-E09 has a minimum predicted frequency of ~1000 Hz.

Experimental evidence shows that, independent of the host fluid, as reorientation time increases, there is a point at which no physical reorientation will take place. The electrostatic torque on the flake is not enough to overcome counter-acting forces such as electrostatic attraction to a surface. Therefore, reorientation will not physically be observed for a PCLC flake in SIT7757 oil, even though a minimum reorientation time of ~120 s is predicted in Fig. 7(d). Reorientation of PCLC flakes in DMS-E09 oil, however, is expected near 1000 Hz with response times <60 s.

For PC, the analytical portion of the model predicts an unrealistic *negative-frequency* independent response for the reorientation time and therefore it is not plotted in Fig. 7(d). When the electric field becomes insufficient to reorient the flake, the analytical model will predict a reorientation time <0. This indicates flake reorientation in a negative direction (to θ ~0°) due to the gravitational torque [13,15].

3.3. Material and Geometry Selection for Increased Electrostatic Torque

Additional modeling was conducted to determine how PCLC flake/fluid host suspensions would behave in microencapsulated geometries resembling electrophoretic (EP)–type devices [18,19,26,27].

Figure 8(a) shows a direct encapsulation type-II cell configuration resembling a typical EP device that consists of a PCLC flake in a low dielectric constant fluid that has been encapsulated into a low dielectric constant film-forming binder. The resulting high electric field is nearly uniform because the dielectric constants and conductivities of all components are similar. For an EP device, a system with low dielectric constants and conductivities is ideal. For a PCLC flake device, however, both high electric field penetration and dielectric mismatch of the fluid and particle are needed for MW reorientation to occur. Therefore, the device pictured in Fig. 8(a) is nonfunctional.

One obvious way to achieve high electrostatic torque in a direct encapsulation type-II configuration is to microencapsulate a low dielectric constant PCLC flake and a moderate dielectric constant fluid host within a film-forming binder with a moderate dielectric constant, as shown in Fig. 8(b). The slightly darker gray region within the capsule and adjacent to the flake denotes a high electric field strength. The strong electric field combined with the mismatch in dielectric constant between the flake and the fluid host leads to a functional, MW driven PCLC flake device. The distortion of the electric field near the poles [i.e., darker areas above and below the capsules shown in Fig. 8(b)], is caused by the abrupt changes in dielectric properties along the path of the electric field between electrodes. At the capsules' closest point to the electrodes, the electric field lines are perpendicular to the capsule and the change in the electric field is the greatest.

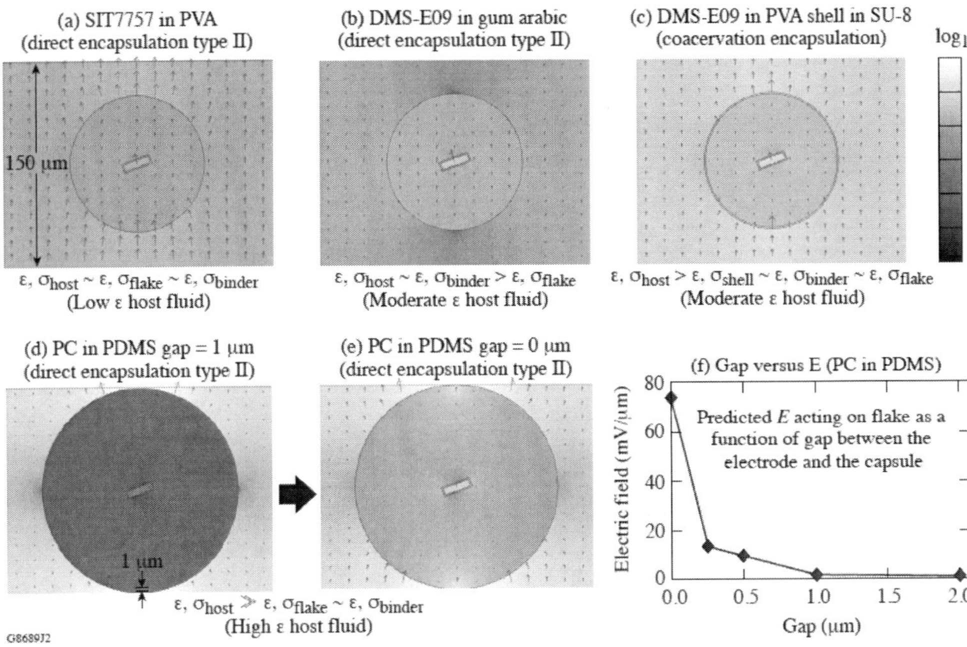

Figure 8. Material and geometry effects of microencapsulation on the electric field at an applied potential of 7.5 V_{rms} for a 150-μm cell gap. The bar scale for the electric field's magnitude on all plots has a range of 0 (black) to 10^6 (white) V/m. (a) PCLC flake suspended in a low dielectric constant host fluid encapsulated in a low dielectric constant binder (e.g., SIT7757 in PVA) shows high electric field magnitude in the host fluid. (b) PCLC flake suspended in a moderate dielectric constant host fluid encapsulated in a moderate dielectric constant binder (e.g., DMS-E09 in Gum Arabic) shows high electric field magnitude in the host fluid. (c) PCLC flake suspended in a moderate dielectric constant host fluid encapsulated in a low dielectric constant capsule, 1-μm-thick capsule wall, dispersed in a moderate dielectric constant binder (e.g., DMS-E09 encapsulated in a PVA capsule dispersed in a SU-8 3050 binder), shows high electric field magnitude in the host fluid. (d) PCLC flake suspended in a high dielectric constant host fluid directly encapsulated in a low dielectric constant binder (e.g., PC emulsified into a PDMS binder) with a 1μm gap between the electrodes and capsule wall shows low electric field strength in the host fluid. (e) PC emulsified into a PDMS binder, as in (d), but now the capsule bridges the gap between the electrodes, resulting in a moderate electric field magnitude in the host fluid. (f) Plot showing the increase in electric field magnitude as the gap between the capsule wall and electrode decreases. The difference in the plots of electric field strength for (a) – (c) are caused by material changes only.

An added level of sophistication that would improve process flexibility is to encapsulate the flake/fluid host suspension in a capsule shell of a different material than the binder prior to dispersing into a film-forming binder. Such a system would create a greater choice in polymer binders. Concerns such as poor capsule formation caused by miscibility of the host fluid in the binder, or a wide distribution of capsule sizes present because of nonuniform mixing, are no longer issues because the capsules are created in a prior process. Figure 8(c) shows this type of configuration (coacervation encapsulation) with the flake/fluid host suspension encapsulated in a capsule shell composed of PVA. For the material set indicated, the polymer binder (SU-8 3050) and host fluid (DMS-E09) are miscible with each other and require the flake/fluid host suspension to be encapsulated before being dispersed in the polymer binder. In Fig. 8(c), this material combination is shown to give a moderately high electric field acting on the flake. As in previous examples, care must be taken when selecting materials. If a capsule shell with too high a dielectric constant is used, it will shield the host fluid and flake from the electric field and result in a nonfunctional device. A further level of sophistication can also be added to customize the capsule shells; additives may be incorporated to promote flake bistability or charge control.

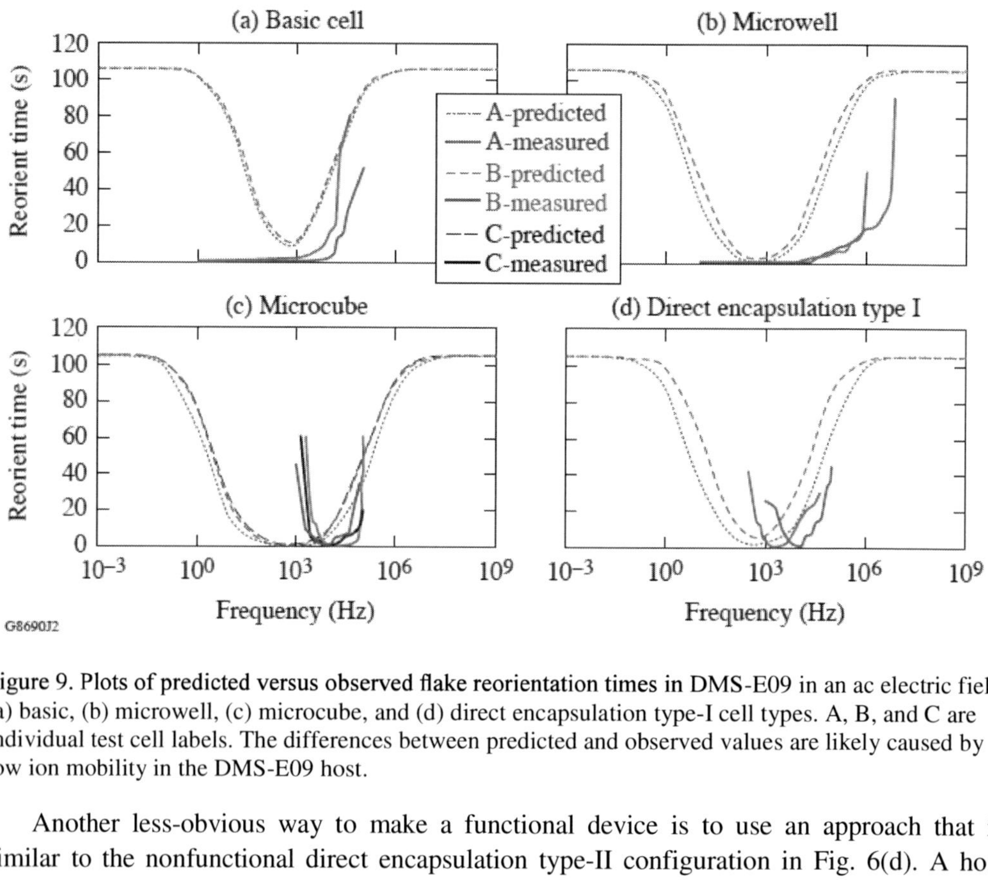

Figure 9. Plots of predicted versus observed flake reorientation times in DMS-E09 in an ac electric field (a) basic, (b) microwell, (c) microcube, and (d) direct encapsulation type-I cell types. A, B, and C are individual test cell labels. The differences between predicted and observed values are likely caused by low ion mobility in the DMS-E09 host.

Another less-obvious way to make a functional device is to use an approach that is similar to the nonfunctional direct encapsulation type-II configuration in Fig. 6(d). A host fluid with higher dielectric constant (PC) can be directly encapsulated into an immiscible low dielectric constant binder (PDMS) and still result in a functional device if the gap between the electrode and capsule wall is very small ($<1\mu$m). As seen in the transition from Figs. 8(d) and 8(e), the electric field can be greatly enhanced within the host fluid if the capsule is in contact with the cell electrode. As shown by the plot in Fig. 8(f), the greater the portion of the cell gap that is filled by the capsule, the higher the magnitude of the electric field acting on the PCLC flake. A gap of 2μm between the capsule and electrodes gives an electric field of $E = 1$ mV/μm, while a gap of 0.5 μm gives E= 15 mV/μm, and a gap of 0 μm gives E = 74 mV/μm.

3.4. MW Behavior in an AC Electric Field

Each device configuration listed in Table 3 was tested in both ac and dc fields. The focus of the modeling reported here is on flake reorientation due to MW behavior in an ac electric field. As predicted by the modeling work shown in Fig. 6 with PC as the host fluid, MW reorientation behavior was observed in basic [Fig. 6(a)] and microwell [Fig. 6(b)] configurations, but not in microcube [Fig. 6(c)] and direct encapsulation type-II [Fig. 6(d)] configurations. No MW re-orientation is either predicted theoretically or observed experimentally for any configuration when SIT7757 is used as the host fluid for PCLC flakes. MW reorientation behavior is both predicted and observed with all six types of encapsulation geometries shown in Fig. 4 when DMS-E09 is used as the host fluid. The moderate dielectric constant of the host fluid permits a significant electric field within the host fluid for all configurations while still providing a reasonable difference in dielectric constants when compared to the PCLC flake. This balance of dielectric constants in the cell gap makes it possible for MW polarization-induced effects to occur for all geometries.

To illustrate the effectiveness of this modeling method, the predicted and observed reorientation times as a function of frequency and geometry for DMS-E09 are compared in Fig. 9. For each type of test cell configuration examined here, the electric field strengths used for predicting reorientation times are matched to those required for making experimental observations. (Drive voltage variations were required for cell tests due to variations in test cell dimensions and dielectric properties. The as-built test cell dimensions varied slightly from the nominal dimensions listed in Table 3 because of manufacturing variations.)

As seen in Figs. 9(a)-9(d), the predicted reorientation times encompass various widths across the frequency-response range. This is a result of the varying electric field magnitudes used in both the model and the experiment. The predicted minimum reorientation times and the widths of the predicted frequency-response curves are both independent of cell configuration and depend only on the electric field acting on the flake. If the predicted reorientation times were plotted with the same electric field strength acting on the flake for all test cell configurations, the curves would be identical.

The dashed lines in Fig. 9(a) show the predicted reorientation times for the basic test cell configuration using drive voltages of 38 and 35 $mV_{rms}/\mu m$. A minimum of ~10 s is predicted near 1 kHz with an increase to a constant value of 105 s at ~1 Hz and ~1 MHz. The experimental data in Fig. 9(a) show a similar trend at higher frequencies, with minimum reorientation times near 1 kHz. Unlike the model predictions, the observed reorientation times remain at constant value of ~1 s below 1 kHz. This is likely due to the host fluid being in contact with the electrodes, since ions have been shown to diffuse out of the ITO and increase the conductivity of the test cell [13].

The predicted reorientation times for the microwell test cell configuration in Fig. 9(b) use drive voltages of 88 and 68 $V_{rms}/\mu m$ and are similar in shape to the predictions for the basic test cell configuration. The experimental data, like those measured for the basic test cell configuration, are similar in shape in the higher frequency regime, but below 1 kHz the observed reorientation times remain at a constant value of ~1 s. These experimental results follow those discussed above for the basic test cell configuration.

The drive voltages used for the microcube test cell predictions in Fig. 9(c) are 240, 181, and 176 $V_{rms}/\mu m$ significantly larger than for the basic and microwell test cell configurations. Unlike the behaviors seen for the basic and microwell test cell configurations, the experimental reorientation times are similar in shape to the predicted values. The frequency for achieving the minimum reorientation time (~10 kHz) is higher than predicted. The observed frequency response range, ~1 to ~100 kHz, over which the flakes reorient, is narrower than predicted. Both the narrower width and symmetric shape of the observed plots are due to the host fluid being insulated from the electrodes by a layer of SU-8 photoresist. Without contact with the electrodes, current flow through the host fluid is negligible, and therefore the conductivity of the host fluid is insignificant. The difference in dielectric constant between the host fluid and flake becomes significant only for inducing the electrostatic torque. This is also likely the reason a larger electric field is needed to achieve low reorientation times comparable to those seen in the basic test cell configuration.

The electric field strengths used for the direct test cell predictions in Fig. 9(d) are 143 $V_{rms}/\mu m$ for both curves. The experimental frequency-response range, ~500 Hz to ~90 kHz, over which the flakes reorient, is narrower than predicted. The experimental reorientation times are similar in shape to the predicted values, but the frequency for the predicted minimum reorientation time is higher (~5 kHz). Like the behavior seen for the microcube configuration, the width, shape, and minimum frequency observed differ from those seen for the basic and microwell configurations because the host fluid is insulated from the electrodes by a layer of SU-8 photoresist.

Cell types with the host fluid in contact with the ITO layer [e.g., basic, microwells, Figs. 9(a) and 9(b)] show a wide frequency range for the minimum MW reorientation time in ac electric fields, even extending into the millihertz regime. Cell types where DMS-E09 is insulated from the ITO layer [e.g., microcubes and direct encapsulation type II, Figs. 9(c) and 9(d)] show MW reorientation in the ac regime and have a minimum reorientation time at a higher frequency than predicted. When the host flake/fluid host suspension is in contact with the ITO layer, ions diffuse out of the ITO and increase the conductivity of the test cell [13]. In a basic cell type, with a host fluid having a high dielectric constant (e.g., PC), the increased ion concentration from the ITO lowers the frequency for the minimum reorientation time. As shown in Fig. 9, the observed frequency for minimum reorientation times in DMS-E09 are slightly higher than predicted. The difference in frequency ranges over which the two host fluids show MW reorientation behavior is likely due to the difference in ion mobility between the two fluids (PC $\sigma = 10^{-4}$ S/m, DMS-E09 $\sigma = 10^{-8}$ S/m).

3.5. EP Behavior in a DC Electric Field

The analytical portion of the model can be modified further to make it possible to predict flake reorientation in a dc electric field [14]. Additionally, EP translation of low-dielectric constant PCLC flakes in a dc electric field is also possible. Unfortunately, the analytical portion of the model used here is not applicable to EP motion so no predictive modeling was done. In a dc electric field with SIT7757 as the host fluid, EP motion has been observed for all geometry types. This observation indicates that the flakes have an intrinsic charge in SIT7757. The direction of translation is toward the positive electrode, which implies a negatively charged particle. Figure 10 shows the observed translation times as a function of applied voltage for different geometries. All configurations follow the same relationship, with the translation time decreasing with increasing electric field strength. This behavior is to be expected because the EP force is directly proportional to the effective electric field [28]. Also, all materials in the cell gap (e.g., SIT7757 and PVA, SU-8 3050) have low dielectric constants and conductivities. Therefore, all configurations have very similar electric field values acting on the PCLC flakes, and only the minimum voltage for initial translation is different for each configuration. EP motion of PCLC flakes in a dc electric field has also been observed in both DMS-E09 and PC as host fluids; however, it is combined with reorientation at low electric field strengths and chaotic electrorotation at higher field strengths.

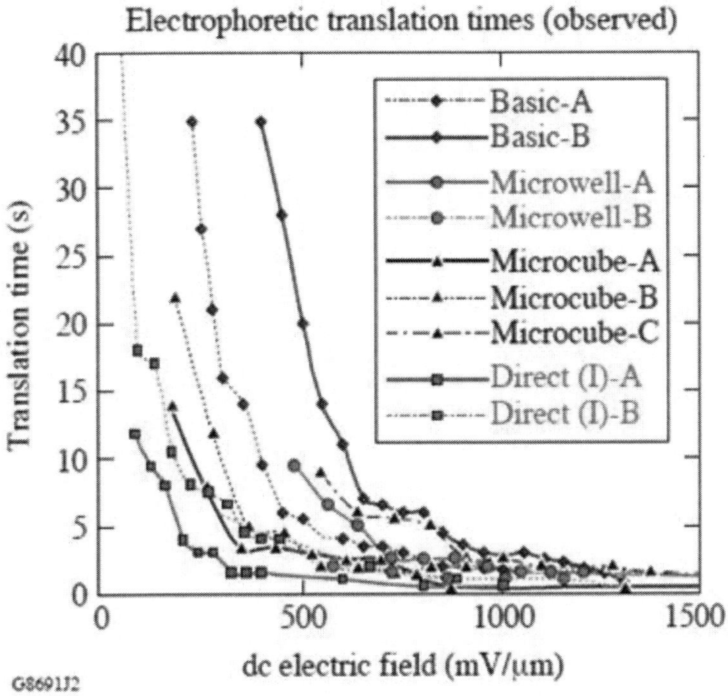

Figure 10. Observed PCLC flake translation times in SIT7757 as a function of applied dc electric field magnitude and test cell geometry. All configurations follow the same relationship, with the translation time decreasing with increasing electric field strength. A, B, and C are individual test cell labels.

3.6. Reverse Drive Mechanism

SIT7757, a low conductivity and low dielectric constant fluid, is found to have increased conductivity resulting from doping. Undoped SIT7757 is hydrophobic and does not show MW reorientation over any frequency range in ac or dc electric fields with any of the test cell configurations. It is known that a small amount of water added to an insulating fluid with a suitable surfactant can raise its conductivity [29]. Initial experiments for fluid doping included the addition of water and NaCl for increased ion mobility through the creation of charge carriers within the fluid. Due to the hydrophobic properties of SIT7757, a surfactant was required to disperse the dopant in the fluid, as neither water nor NaCl is miscible in SIT7757. Water emulsified into a hydrophobic liquid with an appropriate surfactant and sufficient mixing will form micelles with sizes dependent on the concentrations of the water and surfactant. The addition of a salt to the water increases the number of charge carriers in solution. Using this knowledge two doping variations were tested in this experiment:

1. SIT7757 with the surfactant Aerosol OT-100 (1%) and water (0.5%) added.
2. SIT7757 with the surfactant Aerosol OT-100 (1%), water (0.475%) and NaCl (0.025%) added.

These solutions were well mixed and measured for conductivity prior to the addition of PCLC flakes. Both dopants raised the fluid conductivity from 1x 10^{-8} S/m to ~ $4x10^{-8}$ S/m. A difference in the transparency at a macro level between the two fluids was observed; the solution with NaCl (0.025%) added was more transparent than fluid mixture with only water (0.5%) and Aerosol OT-100 (AOT) (1%) added as shown in Fig. 11 [25].

When the solutions were viewed with optical microscopy at 100X magnification, the water and surfactant mixture was observed to have distinct water droplets present in the siloxane fluid, while the similar mixture with NaCl salt added had no visible phase separation. Increasing the magnification to 1000X still did not produce visible phase separation in the second fluid, indicating micelles if present are on the order of nanometers in size as seen in Fig. 12 [25].

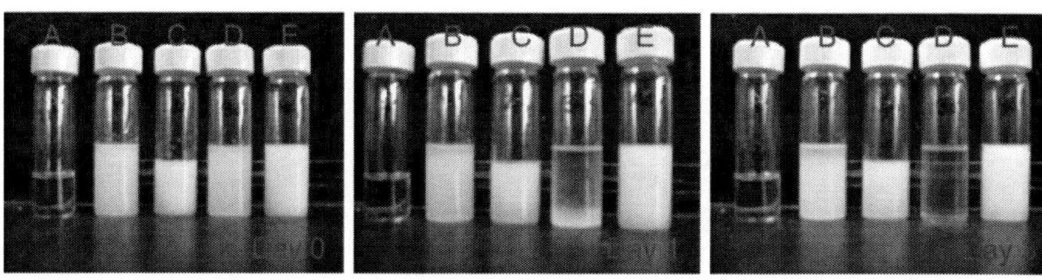

A – (100%) SIT7757
B – (98.5%) SIT7757 + (1%) AOT + (0.5%) H_2O
C – (94%) SIT7757 + (1%) AOT + (5.0%) H_2O
D – (98.5%) SIT7757 + (1%) AOT + (0.475%) H_2O + (0.025%) NaCl
E – (94%) SIT7757 + (1%) AOT + (4.75%) H_2O + (0.25%) NaCl

Figure 11. Images of doped SIT7757 fluid variations aged over a three day period. The formulation of SIT7757 doped with 0.475% water and 0.025% NaCl, labeled as (D), enables the reverse drive mechanism and is shown clearing with time.

PCLC flakes (1%) were added to both solutions and test cells were fabricated by sandwiching PCLC flake/host fluid suspensions between ITO electrode coated substrates (basic configuration). Flake motion greatly differed between the two doping variations. PCLC flakes in the first doping variation of SIT7757 doped with only surfactant and water did not reorient at low voltages (20 - 67 mV_{rms}) and shorted out the cell at higher voltages. PCLC flakes in the second doping variation of SIT7757 doped with AOT, water and NaCl were considered for the remainder of the experiment.

PCLC flakes suspended in SIT7757 with the surfactant, water and NaCl were observed to reorient parallel to the electric field by MW interfacial polarization in both a dc electric field and a low frequency ac field. In a dc electric field, reorientation times were decreased by increasing electric field strength, with response times under a second with less than 400mV/µm. In an ac electric field, MW reorientation parallel to the electric field of PCLC flakes was observed at low frequencies (<40 Hz). Additionally, *unique and novel* electro-optic behavior was observed as the frequency was increased above 40 Hz. At frequencies of

~40 Hz and greater, the MW interfacial polarization of the PCLC flake was inverted and the flake aligned itself perpendicular to the electric field. This reverse reorientation, or turnover, is a bidirectional frequency dependent behavior, allowing for control of the reflective orientation of the PCLC flake surface. A plot of the reorientation times above and below the turnover frequency is shown in Fig 13. This frequency dependent switching behavior in turn enables an on/off switching behavior of PCLC flakes previously unknown and needed to move PCLC flake technology forward to commercial viability for reflective particle displays [30].

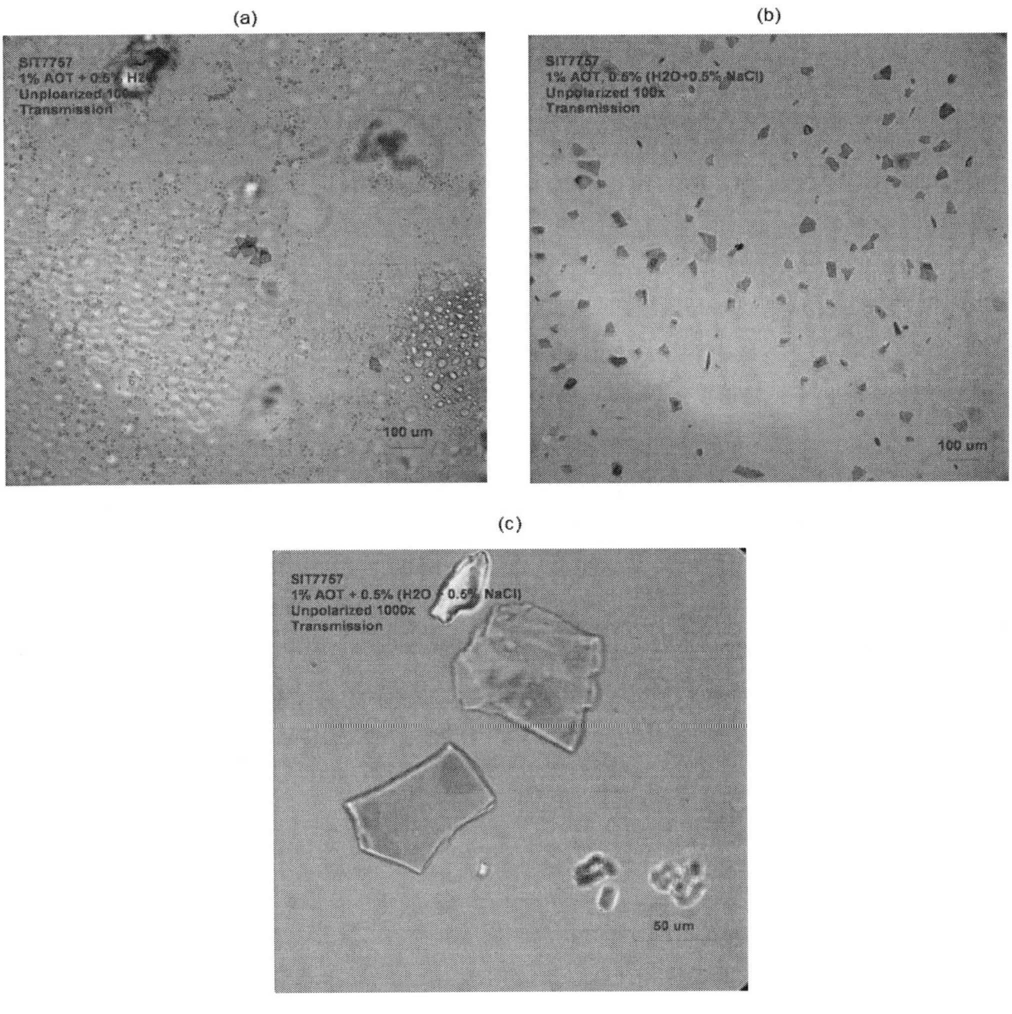

Figure 12. There are significant differences between microscopy images for emulsions with and without NaCl added. (a) SIT7757 host fluid and commercial flakes with AOT (1%), water (0.5%) added, shown at 100x magnification in reflection; the water droplets or micelles are visible and the flakes appear to clump. (b) There are no micelles visible in the image of SIT7757 host fluid and commercial flakes with AOT (1%), water (0.475%) and NaCl (0.025%) added, shown at 100x magnification in reflection. (c) There are no micelles visible in the image of SIT7757 host fluid and commercial flakes with AOT (1%), water (0.475%) and NaCl (0.025%) added, shown at 1000x magnification in reflection.

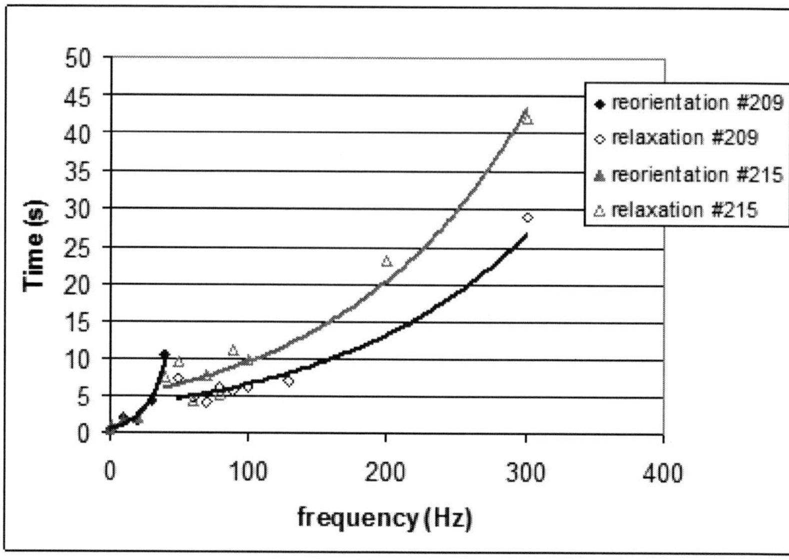

Figure 13. Plot of ac reorientation and driven relaxation times as a function of frequency for PCLC flakes in a SIT host fluid doped with 1% AOT, 0.5% H_2O, and 0.025% NaCl. The flakes align parallel with the electric field < 40 Hz and the flakes align perpendicular with the electric field > 40 Hz. The electric fields were 233 $mV_{rms}/\mu m$, drive current ~ 0.004 μA and cell gaps were ~150μm.

To obtain insight into the mechanism driving this unique electro-optic behavior, the host fluid was isolated from the electrodes in microcube configuration. Microcube test cells were made using a SU-8 epoxy photoresist to encapsulate the flake/host fluid suspension and insulate the ITO electrodes. PCLC flakes in doped SIT7757 exhibited similar motion type in an ac field as in previous trials with basic test cells, but with slower reorientation times and a lower turnover frequency of ~10 Hz (Fig. 14).

In the dc regime, PCLC flakes in the microcube test cells reoriented at a much higher voltage (~ 667 mV/μm) than PCLC flakes in basic test cells. This data indicates that increased conductivity plays an important role in the electro-optical behavior of the PCLC flakes in the basic type cell, even though the conductivity of the doped host fluid is still very low. Also, the fact that reorientation still occurs in the microcube test cells indicate that the difference in dielectric constants between the PCLC flake and host fluid has been changed by doping.

An explanation for the frequency dependent turnover behavior may be found in work done by Saito *et al.* and Jones [31,32]. In an ac electric field, the existence of more than one stable orientation for nonspherical dielectric properties depends upon the relative values of $\varepsilon_2/\varepsilon_1$ and σ_2/σ_1 as well the dimensions of the particles (PCLC flakes), where ε_1, σ_1 are the doped host fluid dielectric properties and ε_2, σ_2 are the PCLC flake dielectric properties. Frequency dependent turnover behavior is guaranteed if either (i) $\varepsilon_2/\varepsilon_1 < 1 < \sigma_2/\sigma_1$ or (ii) $\sigma_2/\sigma_1 < 1 < \varepsilon_2/\varepsilon_1$ [30]. Measured values for the dielectric constant of the doped SIT7757 are difficult to obtain, however a qualitative comparison can prove insightful. Before doping of the host fluid, the dielectric properties of the PCLC flakes and the SIT7757 are nearly equal, σ_1~σ_2 and ε_1~ε_2. It is reasonable to assume that the conductivity of the host fluid, σ_1, increases due to doping. In addition, the acryl functional groups attached to the cyclic siloxane molecules on the PCLC material makes the PCLC flakes more hydrophilic than SIT7757. This difference in hydrophobicity leads to the assumption that adsorption of water and/or ions occur on the surface the flakes, increasing the dielectric constant of the PCLC flakes ε_2. This assumption is supported by data above for the microcube cells, which indicates an increased difference between the dielectric constants of the host fluid and flake. If both assumptions are true, then the condition for turnover behavior of nonspherical dielectric particles has been met, $\sigma_2/\sigma_1 < 1 < \varepsilon_2/\varepsilon_1$.

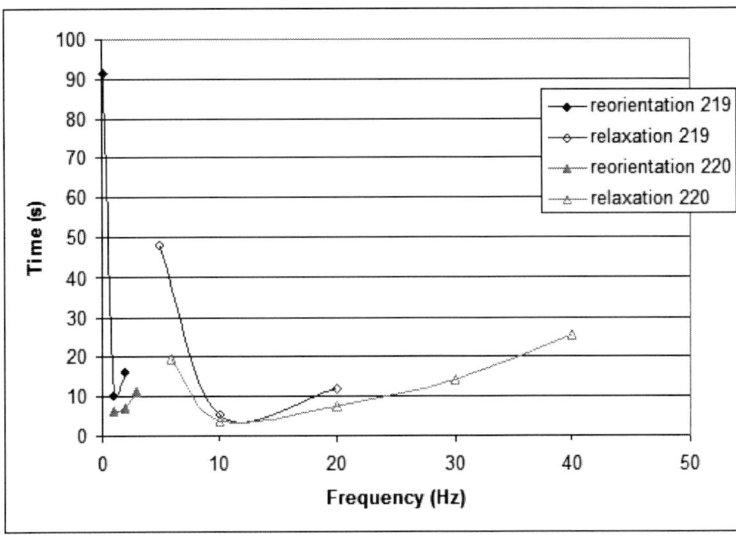

Figure 14. The ac reorientation and driven relaxation times for PCLC flakes in a doped SIT7757 host fluid in a microcube test cell are a function of frequency. The turnover frequency is lower than for a basic test cell and the reorientation times are slower. Electric fields applied are ~267 $mV_{rms}/\mu m$, drive current ~0.004 μA and the turnover frequency is ~5 Hz.

CONCLUSION

By using a combination of COMSOL MULTIPHYSICS and an analytical model presented in this chapter, we can predict both the resultant electric field for any given device geometry and its effect on PCLC flake motion caused by MW polarization. The effect of the surrounding environment on the PCLC flake's electro-optical behavior depends greatly on both the materials used and the type of microencapsulation geometry. The MW reorientation behavior of PCLC flakes is directly proportional to both the effective electric field acting on the flake and the difference in dielectric constants and conductivities between the fluid host and flake. The optimal cell geometry for MW reorientation will have a balance of uniform dielectric constants and conductivities along the path of the electric field and a mismatch in dielectric constants between the host fluid and flake. With proper matching of device geometry and material properties, it is possible to encapsulate a range of host fluids and shaped PCLC flakes while maintaining MW reorientation capability for the flake/fluid host suspension. In addition, good control has been demonstrated over EP-type motion of microencapsulated PCLC flakes in a host fluid with low dielectric constant and conductivity.

It was also shown that by doping a low dielectric property host fluid it is possible to induce MW interfacial polarization of PCLC flakes and reorient them parallel to an electric field. An anionic surfactant was used to incorporate water and salt into a siloxane fluid, where the resulting mixture had higher dielectric properties and was used as a host fluid for PCLC flakes. The addition of salt was a critical part to doping success, as without it, the surfactant was unable to fully disperse the water into the siloxane fluid. The dopants also modified the dielectric properties of the flake/host fluid suspension such that turnover behavior became possible at a critical electric field frequency. This frequency dependent switching behavior in turn enables an on/off switching behavior of PCLC flakes previously unknown and needed to move PCLC flake technology forward to commercial viability for reflective particle displays. The new extensions in understanding presented in this chapter continue PCLC flakes down the path to a viable particle display technology.

ACKNOWLEDGMENTS

The work presented here was done at the University of Rochester, Laboratory for Laser Energetics in Rochester, NY, USA. The author would like to thank Prof. Stephen Jacobs for his guidance and supervision through this project, Mr. Kenneth Marshall, Dr. Anka Trajkovska-Petkoska and Dr. Tanya Kosc for their constructive advice and help. Author, also acknowledges the Laboratory for Laser Energetics for generous Horton Fellowship support.

REFERENCES

[1] Korenic, E. M.; Jacobs, S. D.; Faris, S. M.; Li, L. *Mol. Cryst. Liq. Cryst.* 1998, *317*, 197.

[2] Trajkovska-Petkoska, A.; Varshneya, R.; Kosc, T. Z.; Marshall, K. L.; Jacobs, S. D. *Adv. Funct. Mater.* 2005, *15*, 217.

[3] Trajkovska-Petkoska, A.; Jacobs, S. D.; Kosc, T. Z.; Marshall, K. L. Manufacture of Regularly Shaped Polymer Cholesteric Liquid Crystal Flakes with a Mechanically Flexible Mold, U.S. Pat. 7,238,316 B2, 2007.

[4] Marshall, K. L.; Kosc, T. Z.; Trajkovska-Petkoska, A.; Kimball, E.; Jacobs, S. D. presented at the Fourth Annual Flexible Microelectronics and Displays Conference, Phoenix, AZ, 1-3 February 2005 (unpublished).

[5] Marshall, K. L.; Hasman, K.; Leitch, M.; Cox, G.; Kosc, T. Z.; Trajkovska- Petkoska, A.; Jacobs, S. D. *SID Int. Symp. Digest Tech. Papers* 2007, *38,* 1741.

[6] Marshall, K. L.; Kosc, T. Z.; Jacobs, S. D.; Faris, S. M.; Li, L. *U.S. Patent No. 6,665,042 B1,* 2003.

[7] Kosc, T. Z. Ph.D. thesis, University of Rochester, 2003.

[8] Kosc, T. Z.; Marshall, K. L.; Jacobs, S. D.; Lambropoulos, J. C.; Faris, S. M. *Appl. Opt.* 2002, *41*, 5362.

[9] Kosc, T. Z.; Marshall, K. L.; Jacobs, S. D. *U.S. Patent No. 6,829,075 B1*, 2003.

[10] Kosc, T. Z.; Marshall, K. L.; Jacobs, S. D. *U.S. Patent No. 7,042,617,* 2006.

[11] Marshall, K. L.; Kimball, E.; McNamara, S.; Kosc, T. Z.; Trajkovska-Petkoska, A.; Jacobs, S. D. *Liquid Crystals VIII*, edited by I.-C. Khoo, SPIE, Bellingham, WA, 2004, Vol. 5518, p. 170.
[12] Kosc, T. Z.; Marshall, K. L.; Jacobs, S. D.; Lambropoulos, J. C. *J. Appl. Phys.* 2005, *98*, 013509.
[13] Trajkovska-Petkoska, A. Ph.D. thesis, University of Rochester, 2007.
[14] Trajkovska Petkoska, A.; Kosc, T. Z.; Marshall, K. L.; Hasman, K.; Jacobs, S. D. *J. Appl. Phys.* 2008, *103,* 094907.
[15] Cox, G. P.; Marshall, K. L.; Lambropoulos, J. C.; Leitch, M.; Fromen, C.; Jacobs, S. D. *LLE Review Quarterly Report* 2009, *118*, 80.
[16] E Ink Corp., Electronic Paper Displays, E Ink Corporation, Cambridge, MA 02138.
[17] Chen, Y.; Au, J.; Kazlas, P.; Ritenour, A.; Gates, H.; McCreary, M. *Nature* (London) 2003, *423,* 136.
[18] Comiskey, B.; Albert, J. D.; Yoshizawa, H.; Jacobson, J. *Nature* (London) 1998, *394,* 253.
[19] Guo, H. L.; Zhao, X. P. *Opt. Mater.* 2004, *26*, 297.
[20] Shiyanovskaya, I.; Khan, A.; Green, S.; Magyar, G.; Doane, J. W. *SID Int. Symp. Digest Tech. Papers* 2005, *36,* 1556.
[21] Stephenson, S. W.; Johnson, D. M.; Kilburn, J. I.; Mi, X.-D.; Rankin, C. M.; Capurso, R. G. *SID Int. Symp. Digest Tech. Papers* 2004, *35*, 774.
[22] Szep, J. Electronic Paper Edging Toward Reality, Computerworld-Mobile & Wireless 2007.
[23] Kosc, T. Z. *Opt. Photonics News* 2005, *16*, 18.
[24] Cox, G. P. Microencapsulation effects on the electro-optic behavior of polymer cholesteric liquid crystal flakes, PhD thesis, University of Rochester, 2009.
[25] Trajkovska Petkoska, A. *Enhanced Electro-optical Behavior of PCLC Flakes in Host Fluids,* PhD Thesis, University of Rochester, Rochester, NY, 2007.
[26] Song, J. K.; Choi, H. J.; Chin, I. *J. Microencapsul.* 2007, *24*, 11.
[27] Song, J. K.; Kang, H. C.; Kim, K. S.; Chin, I.-J. *Mol. Cryst. Liq. Cryst.* 2007, *464,* 263.
[28] Morrison, I. D.; Ross, S. *Colloidal Dispersions: Suspensions, Emulsions, and Foams*; 2002, Wiley-Interscience: New York.
[29] Morrison, I. D. Coll. Sur. A: Physicochemical and Engineering Aspects 1993, 71, 1-37.
[30] Cox, G. P.; Fromen, C. A.; Marshall, K. L.; Jacobs, S. D. *U.S. Patent No. 8,293,135* 2012.
[31] Saito, M.; Schwan, H. P.; Schwarz, G. *Biophys. J.* 1996, 6.
[32] Jones, T. B. *Electromechanics of particles*; 1995, Cambridge University Press: New York.

In: New Developments in Liquid Crystals Research
Editors: A. T. Petkoska and A. T. Broach
ISBN: 978-1-63482-456-9

Chapter 4

EFFECT OF NANOMATERIALS ON PHYSICAL PROPERTIES OF LIQUID CRYSTALS

V. Manjuladevi*[*] *and R. K. Gupta
Department of physics, BITS Pilani, Rajasthan, India

ABSTRACT

Liquid crystals are the fourth state of matter whose order lies between that of 3-dimensionally ordered solids and completely disordered fluids. The physical parameters such as threshold voltage, dielectric anisotropy, and elastic constants are the key factors in designing liquid crystal display devices. In this chapter, we have reviewed the effect of nanomaterials viz. carbon nanotubes, quantum dots, and nanoparticles and their derivatives in liquid crystal-nanocomposites on the electro-optic switching, threshold voltage, dielectric anisotropy and elastic constants of the nanocomposites.

Keywords: Liquid crystals, dielectric anisotropy, elastic constant, carbon nanotubes, quantum dots, order parameter

1. INTRODUCTION

From past two decades, the research and development in the fields of nanomaterials and nanostructures have made tremendous impacts on every branch of science and technology. Nanoscale materials with physical dimensions of 1 to 100 nanometers (nm), featuring unique properties different from the bulk, have revolutionized many fields including catalysis, sensors, photonics, energy, computing and medicine. The materials exhibit extraordinary properties at the nanometer length-scale. The limitation to the material properties can be addressed by the reduction of dimension and exploring the science and technology at the nanometer length-scale. Various systems can be doped with nanomaterials to improve its physicochemical properties. Liquid crystal display (LCD) is the heart of 80% of appliances

[*] email: manjula@pilani.bits-pilani.ac.in

starting from household television to medical research. Enormous amount of effort is being invested to upgrade the field of liquid display to a matured level. The incorporation of nanomaterials to liquid crystal (LC) matrix can lead to high performance LCDs and thus creates an avenue for research and development in the field of nanomaterials doped LC systems. Along with a revolution in the field of nanomaterials, the field of liquid crystals (LCs) is a particularly vibrant area wherein the use of nanomaterials with different size, shape, core material, and surface functionality can prove to be beneficial. Liquid crystal nanoscience as a subject can be approached from two major perspectives [1]. One perspective would focus on using reconfigurable liquid crystals either for template-assisted syntheses of nanomaterials [1-4] or using these self-organizing anisotropic fluids to assemble and manipulate nanoscale [5-7] materials. The second one would encompass approaches where nanomaterials such as nanoparticles (NPs) have effects on liquid crystals based on very fundamental interactions between liquid crystal host molecules and nanomaterials as dopants [8-10].

Liquid crystals (LCs) are well known organic materials used in many electronic devices such as flat panel liquid crystal displays, sensors and light modulators [11,12]. Liquid crystals are spontaneously self organized anisotropic fluids [13,14]. They exhibit optical, dielectric and magnetic anisotropy similar to solids; and flow like liquids. Such interesting and unique properties of LCs made it possible to fabricate liquid crystal displays which possess numerous advantages over the conventional cathode ray tube (CRT) displays. Based on the symmetry of the medium, LCs are mainly classified as: nematic (N), smectic (Sm) and cholesteric (N*). The simplest of the LC phases is nematic (N), in which the long axis of the molecules on an average orient along a preferred direction, called the director ($\hat{n}$). The director, $\hat{n}$ is a dimensionless apolar unit vector i.e., $\hat{n}$ and $-\hat{n}$ are physically equivalent [14]. Even when the molecules are made of highly polar end groups, the medium is still apolar as the molecules orient such that there is no net polarization in the medium. Liquid crystal displays (LCD) have now become the most promising technology for a wide range of applications starting from simple black and white pocket calculators to high end full color television screens. Most commonly used LCs in these displays are in N phase. All of these applications on nematic displays work on the principle of Freedericksz transition, a collective reorientation of the LC molecules possessing dielectric anisotropy along the direction of applied electric field [13].

Over the years, a number of modes of LCD have been predicted and realized [15,16]. They are guest-host (GH), dynamic scattering (DS), twisted nematic (TN), hybrid aligned nematic (HAN), super twisted nematic (STN), optically compensated bend (OCB), in-plane switching (IPS), and vertical alignment (VA) modes. Among all these modes, the TN mode [17,18] is popular in display industry due to its advantages of low operation voltage, high contrast ratio and nearly achromatic transmission. The performance of all the LCD modes is influenced by the alignment layer and the LC employed. In thin film transistor (TFT) LCD, the LC cell should be driven by ac voltage. On application of dc electric field to LC samples with low resistivity, the ion charges migrate towards the enclosing surface leading to surface polarization. This screening effect due to ion charges reduces the effect of applied dc voltage. The ion-charge effect gives rise to flickering and image sticking, which can degrade the quality of LCD. Hence, in order to avoid flickering and surface polarization, materials with high resistivity [19] are being used for LCD manufacturing.

Most commonly known and easily synthesizable LCs such as cyanobiphenyls possess low resistivity, and hence cannot be used in TFT-LCD industry in their pure form. However, the physical properties of such LC materials can be tuned by doping appropriately with nanomaterials such as metal nanoparticles, nanorods and carbon nanotubes. There are some reports [20-23] available in literature which indicate that on doping with nanoparticles, the parameters of liquid crystal switching can be altered. Recently [24,25] we have investigated the effect of a dopant, octadecylamine functionalized carbon nanotube (ODACNT) on the electro-optic response and dielectric measurements of a host LC. The threshold electric field of ODACNT doped LC was less as compared to that of pure LC whereas the dielectric anisotropy is enhanced. The tunneling current images obtained using atomic force microscope in spreading resistance mode indicate presence of ODACNT in close proximity of alignment layer. These results clearly indicated that ODACNT is influencing the nematic anchoring as well as the ordering in the bulk due to π–π interaction between LC, ODACNT and the alignment layer.

There are many groups concentrating on their research on studying the effect of doping spherical nanoparticles [8-10, 26-36] with liquid crystals and studying the effect of dopant nanoparticles on the LC alignment, switching time etc. Kopcansky et al. [37] have doped a liquid crystal with magnetite nanoparticles of different shapes. They have reported that on doping the LC with rod like and chain like particles, a rigid anchoring with parallel initial orientation between magnetic moments of magnetic particle and the nematic director is observed. The critical magnetic field for switching the liquid crystal reduces on doping with spherical magnetic nanoparticles. On doping the LC with chain-like magnetite nanoparticles a further reduction of critical magnetic field compared to that of spherical nanoparticles is observed. In 2011, D. Zhao et al. have reported [38] that liquid crystals can be aligned without any external aligning agent on doping with nickel nanoparticles of spherical and bowl shapes. They have observed homeotropic alignment of the liquid crystal on doping with nickel nanospheres and a planar alignment on doping with nickel nanobowls. They attribute the alignment of liquid crystals to the adsorption of doped nickel nanoparticles on the inner surfaces of ITO substrate. In the year 2012, Podoliak et al. have reported that liquid crystals doped with functionalized gold nanoparticles exhibit high optical nonlinearity [26]. The functional group tetra octylammonium bromide plays a crucial role in increasing the pre-tilt angle and thereby, reducing the anchoring strength which leads to stronger modulation of both molecular reorientation and refractive index of the LC system.

Basu and Innacchione [27] have reported that the assembly of CdS quantum dots in a LC matrix into nanoscale macroscopic chain like configurations. They have observed that doping a nematic LC with CdS quantum dots led to longer relaxation times. Based on the dielectric measurements as well as electro-optic measurements, they have demonstrated considerable changes in switching time and average dielectric response of LC by the presence of long anisotropic pearl-necklace strands made of quantum dots. Hegmann et al [1] have incorporated a series of CdSe and CdTe quantum dots of various sizes and multiple functionalities in nematic, smectic and columnar LC materials. They have observed lower threshold voltages and the altered dielectric permittivities. The results obtained from electro-optic measurements using the CdTe quantum dots as additives indicated that the QDs with shorter alkyl chain length in the capping layer induced more pronounced effects regarding the electro-optic parameters of nematic LC in comparison to other QDs and even gold nanoparticles at the same size regime.

Most recently S Kumar et al. have shown that gold nanorods (GNRs) can be easily inserted into the supramolecular assembly of discotic liquid crystals (DLCs) along the director without disturbing their mesomorphism [35]. GNRs embedded discotic nanoribbons can be useful for constructing devices like thin film transistors and sensors. Such Systems can be prepared by simple solution processing of GNR–DLC nanocomposites.

S Kumar et al. have dispersed cadmium selenide quantum dots (QDs) in a discotic liquid crystal columnar matrix [39]. Thermo-physical properties of these liquid crystal nanocomposites were investigated by UV-Vis spectroscopy, photoluminescence spectroscopy, differential scanning calorimetry, polarizing optical microscopy, DC conductivity and small angle X-ray diffraction. The composites exhibit enhancement in the DC conductivity when compared with pure discotic liquid crystal. In 2012, Malik et al. have carried out the electro-optic and dielectric measurements on ferroelectric liquid crystal mixture doped with cadmium sulphide nanorods [36]. They report that doping of CdS nanorods increases the spontaneous polarization and response time. This increase is attributed to large dipole-dipole interaction and increase in anchoring energies existing between nanorods and FLC molecules.

2. EXPERIMENTAL

2.1. Liquid Crystal (LC) Cell Fabrication

In an unaligned LC sample, the nematic director, $\hat{n}$ varies from point to point in space. For measuring the electro-optic switching behavior, we need a mono-domain sample of the liquid crystal in which the director $\hat{n}$ is uniform throughout the sample. Generally, either a planar or homeotropic alignment is used. A planar cell is one in which the director is unidirectional and parallel to the enclosing surfaces (Figure 1(a)). On the other hand in a homeotropically aligned sample, the director is perpendicular to the enclosing surfaces (Figure(1)b).

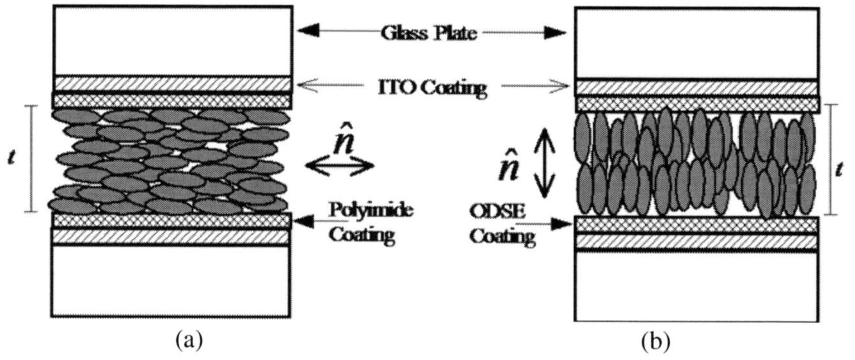

Figure 1. Schematic diagram of the (a) homogeneous (planar) and (b) homeotropic alignment of the liquid crystal (LC) molecules, where $\hat{n}$ represents the director.

Even a small dust particle or ionic impurity can degrade the quality of LC cells, hence they have to be fabricated in a *clean room*. For electro-optic switching measurements, the transmitted intensity of a LC cell is measured as a function of applied voltage. Hence, the cell should be prepared on transparent electrodes such as indium tin oxide coated glass (ITO) plates. In order to obtain planar alignment, ITO plates are coated with polyimide (PI) and then cured at curing temperature of the polyimide for a specific time in a hot air oven. The cured plates are rubbed uni-directionally using a soft cloth (or tissue paper). The cell is prepared such that the rubbing directions of both the enclosing plates are parallel to each other. This kind of processing yields planar geometry. For TN mode, the processed plates are placed such that the rubbing directions of both the enclosing plates are perpendicular to each other. A little gap is left at both the ends for the electrical connections. The sides are sealed using a special kind of glue viz. epoxy glue meant for LCD which does not react with liquid crystals. The thickness of the cell is fixed using either glass beads or Mylar spacers. The sealed cell is cured for about an hour at 150 °C. A transverse electric field is applied in case of TN and planar modes. For homeotropic alignment of the molecules, the ITO plates are coated with a surfactant like octadecyl triethoxy silane (ODSE) and cured at 150 °C for about an hour. ODSE is a long chained molecule which is amphiphilic in nature having a polar group and an aliphatic chain. The polar end group is attracted to the surface of the ITO plate and the long aliphatic chain interacts with the alkyl chain of the liquid crystal molecules thus giving rise to perpendicular orientation of the director with respect to the glass plate.

2.2. Preparation of LC-Nanomaterial Composites

LC- nanomaterial dispersions are prepared by dispersing the LCs and nanomaterials in suitable solvents such as chloroform, ethanol or toluene and then by ultrasonicating the mixture for 2-3 hours.

2.3. Electro-Optic Switching Characterization

The electro-optical switching of the LC-system can be studied using an experimental setup as shown in Figure 2. An optical chopper (C) is used to convert optical beam into an alternating signal of a particular frequency.

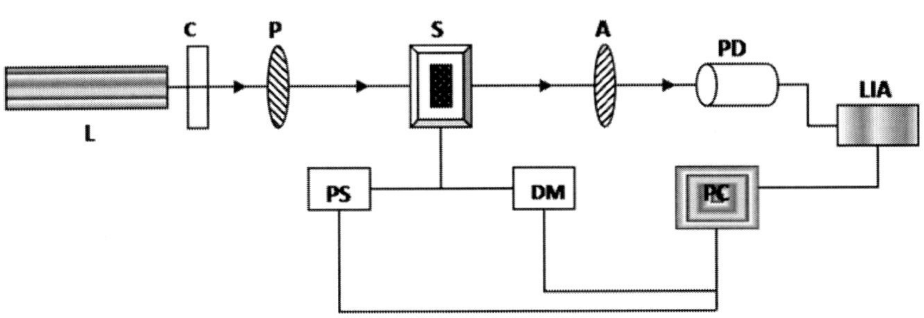

Figure 2. Schematic representation of the electro-optical switching setup: L-Laser, C-Chopper, P-Polarizer, S- LC cell mounted in hot-stage, A-Analyzer, PD-Photo-diode, LIA-Lock-in amplifier, PC-Personal Computer, DM-Digital Multimeter, PS-Power Supply.

By this way fluctuations caused by stray light can be eliminated by a phase sensitive measurement. A sensitive photodiode (PD) is used to precisely measure the transmitted intensity from the LC- and its nanocomposite enclosed in a LC cell kept between the crossed polarizers (P & A). Using sensitive photodiode in conjunction with Lock-in amplifier (LIA) ensures that the measured intensity at a particular frequency is just due to the cell and not due to any other sources.

2.4. Measurement of Threshold Voltage

LC and its nanocomposites are filled into the fabricated cells at appropriate temperatures using capillary action. The prepared samples are mounted in a hot-stage and the textures of these LC and nanomaterial dispersions are observed under a polarized optical microscope, and their images are recorded using a digital camera. Their phase sequence are found out using the optical microscopy and are compared with the Differential Scanning Calorimetry results. The Freedericksz transition voltage is measured using the electro-optic setup. When an electric field is applied to a cell in planar geometry kept between crossed polarizers, the LC molecules as well as nanomaterials will try to reorient themselves along the direction of applied field (Figure 3). This reorientation starts occurring at a certain threshold voltage when the field strength is greater than the surface anchoring strength. This voltage is called the Freedericksz threshold voltage. The threshold voltage is given by $V_{th}=\pi[K_{11}/(\varepsilon_0\Delta\varepsilon)]^{1/2}$ where K_{11} is the splay elastic constant (Figure 3), ε_0 is the vacuum dielectric constant and $\Delta\varepsilon$, dielectric anisotropy of the nematic LC. At high field, all the LC molecules and hence most of nanomaterials will align along the direction of electric field.

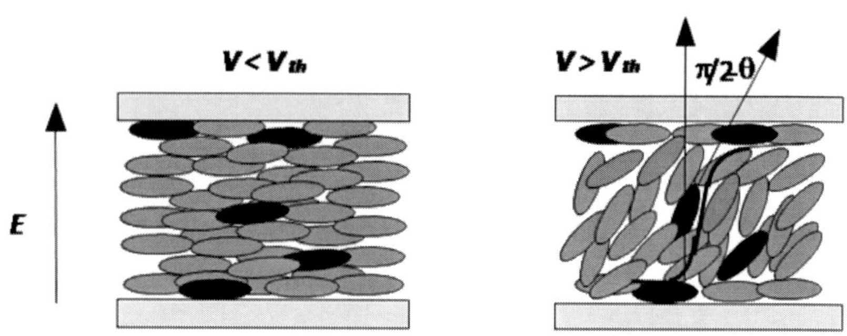

Figure 3. Schematic representation of Freedericksz transition: director configuration for (a) $V<V_{th}$ (b) $V>V_{th}$. The gray and dark ellipsoids represent LC molecules and nanotubes, respectively.

The rubbing direction of the LC cell is treated as the optic axis. The cell is placed between crossed polarizers with the optic axis at an angle of 45° with respect to the polarizer. The transmitted intensity is monitored using the electro-optic setup. The intensity is proportional to the optical phase difference of the sample, $\Delta\phi=2\pi\overline{\Delta\mu}\,t/\lambda$, where $\overline{\Delta\mu}$ is the average value of the birefringence, t is the sample thickness and λ the wavelength of incident light. The distortion of the orientation in the medium increases with increasing applied voltage leading to a decrease in the average value of the birefringence. With increase in the applied voltage, the rate of variation of transmitted intensity decreases. At high values of applied voltages (typically ≥ 5V), the intensity profile flattens out as the average value of the birefringence $\overline{\Delta\mu}$ is negligible. Hence, using this method the absolute value of optical phase difference is estimated. The birefringence, and hence the order parameter of the nematic phase can be calculated. The operating voltage of a LC cell is the voltage beyond which no significant change in transmitted intensity takes place. The electro-optic response of these cells in nematic phase can be recorded on application of ac (square) signal. The transmitted intensity as a function of time can be recorded at different voltages. Optical transmission upon electrical switching in voltage ON and OFF state can be recorded. The rise and decay times are measured from the transmittance variation between 10% to 90% and 90% to 10%, respectively.

2.5. Dielectric Constant and Conductance Measurements

Due to the anisotropic nature of the LC molecule, the nematic phase exhibits dielectric anisotropy, where components parallel and perpendicular to the molecular long axis are $\varepsilon_{\parallel}$ and $\varepsilon_{\perp}$ respectively. The ionic impurities in the LC samples can contribute to a finite resistance of the sample cell. Therefore an equivalent electrical circuit of a liquid crystal cell (Figure 4) is considered to be a capacitance in parallel with resistance [40, 41].

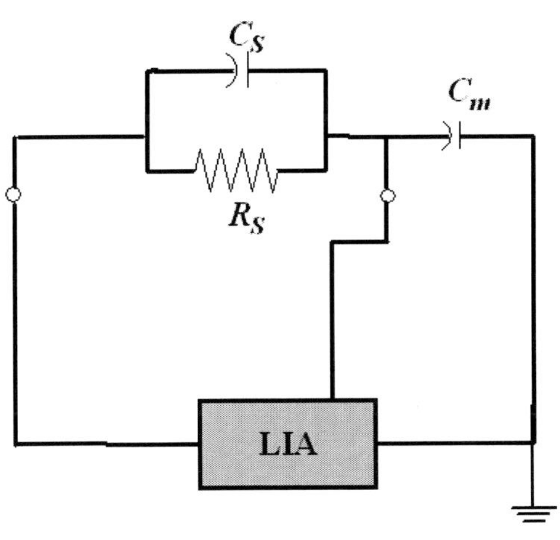

Figure 4. Equivalent circuit used for the impedance measurement of the liquid crystal cell. C_S and R_S are the sample capacitance and resistance, respectively.

Apart from the electro-optic switching measurements, the dielectric and conductance measurements as a function of temperature on the various LC-nanomaterial dispersions are carried out using the setup shown in Figure 4. The dielectric and conductance measurements reveal the physical parameters which are responsible for the changes observed in threshold voltage and response time.

A lock-in amplifier (LIA, SRS, model 830) is used to apply electric field to the sample for performing the dielectric measurements. The capacitance of the empty cell is measured before filling the sample. The ratio of the capacitance of the sample filled cell to the empty cell capacitance gives the dielectric constant of the sample.

For measuring the $\varepsilon_{\parallel}$ component, homeotropic cells is used, wherein the long axis of the LC molecules are perpendicular to the ITO substrate and a transverse electric field parallel to the long axis of molecules is applied. The dielectric constant of each sample is recorded as a function of temperature at a frequency of 4.1 kHz. The data are also recorded as a function of voltage at a given temperature.

2.6. Liquid Crystals and Nanomaterials

The composites of LCs and nanomaterials are formed using the materials listed below (Figure 5):

- E7 (Merck): Mixture of 4 LC compounds: 51% 5CB, 25% 7CB, 16% 8OCB, 8% 5CT (5CB: n-pentyl cyanobiphenyl; 7CB: n-heptyl cyanobiphenyl; 8OCB: n-octyloxy cyanobiphenyl; 5CT: n-pentyl cyanoterphenyl)
- 8CB (Sigma Aldrich) : n-octyl cyanobiphenyl
- 5PCH (Sigma Aldrich): 4-trans-pentylcyclohexylcyanobenzene
- AFCNT(Carbon Solutions Inc.): Acid functionalized Single wall carbon nanotube
- ODA-SWCNT(Carbon Solutions Inc.): Octadecylamine functionalized Single wall carbon nanotube
- CdSe QDs: Octadecyl amine capped Cadmium Selenide Quantum Dots

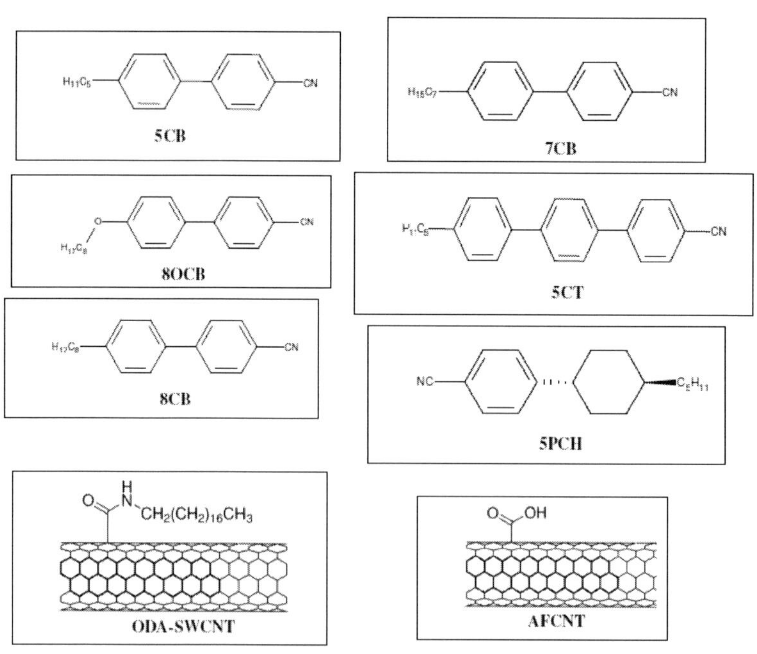

Figure 5. Chemical structures of the liquid crystals and the functionalized single-walled carbon nanotubes.

Although carbon nanotubes (CNTs) exhibit similar shape anisotropy as compared to a calamitic liquid crystal molecule viz. rod shaped, the CNTs can possess larger length to width (*L/W*) aspect ratio. Due to the shape compatibility, the CNTs can be doped in the LC matrix. The *L/W* ratio and the organic functionalization of the CNTs can play a key role in altering the physical properties of the LC medium [24, 42-45]. In 2013, [46] Lim et.al have reported a decrement in electro-optic response time due to the incorporation of thin multi-walled carbon nanotubes in a negative dielectric anisotropic liquid crystal material filled in a patterned vertically aligned cell. Recently, Lee et al [47] have coated the substrates of LC cells with a mixture of multiwall CNT doped polyimide (PI). The sample coated with the PI-CNT mixture showed faster decay time and greater splay elastic constant compared to that coated with the pure PI. Though, there are many studies on determination of elastic constants of calamitic liquid crystals but there is hardly any report on a systematic study of elastic constant of calamitic liquid crystals doped with single - walled carbon nanotubes (SWCNT). Octadecylamine functionalized SWCNTs are soluble in many organic solvents and it is found to be dispersible in liquid crystals at low concentration [24]. Therefore, a nanocomposite of n-octyl cyanobiphenyl and ODA-SWCNTs are prepared and the effect of such doping on the physical properties as a function of temperature was studied.

3. RESULTS AND DISCUSSION

3.1. E7 and Functionalized SWCNTs Systems

A dispersion of ~ 0.05 wt% acid functionalized SWCNTs (AFCNT) was prepared in a commonly studied LC, E7. The thickness of the planar aligned cells was ~ 4.5 μm. All the optical and dielectric measurements were carried out at room temperature (~24 °C).

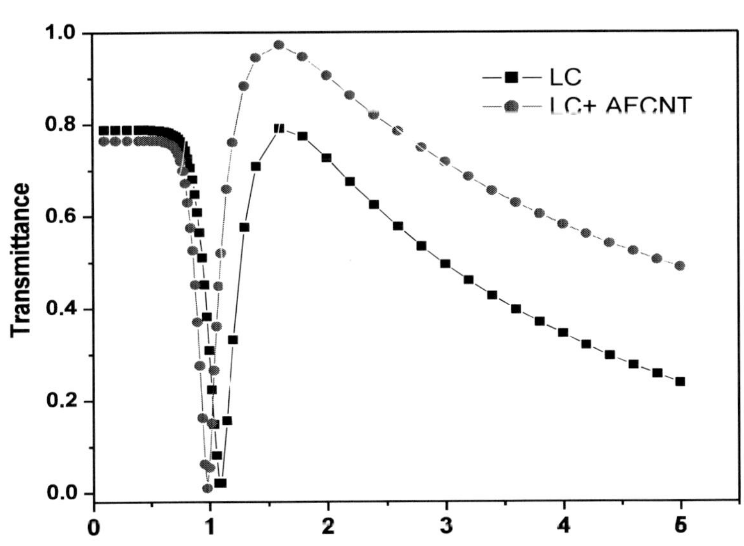

Figure 6. Transmittance of pure LC and AFCNT doped LC (E7) as a function of applied voltage at 2.1 kHz.

The transmittance as a function of applied voltage at 2.1 kHz is shown in Figure 6. Interestingly, the V_{th} is found to be ~ 0.7 V for pure LC as well as AFCNT doped LC. Using the transmitted intensity data, the value of birefringence (Δμ) at 24°C is estimated [48]. The value of Δμ is found to be ~ 0.22 for pure as well as AFCNT doped LC which is in good agreement with previous measurements on pure LC [49, 50]. The dielectric constant (ε) as a function of applied voltage is shown in Figure 7.

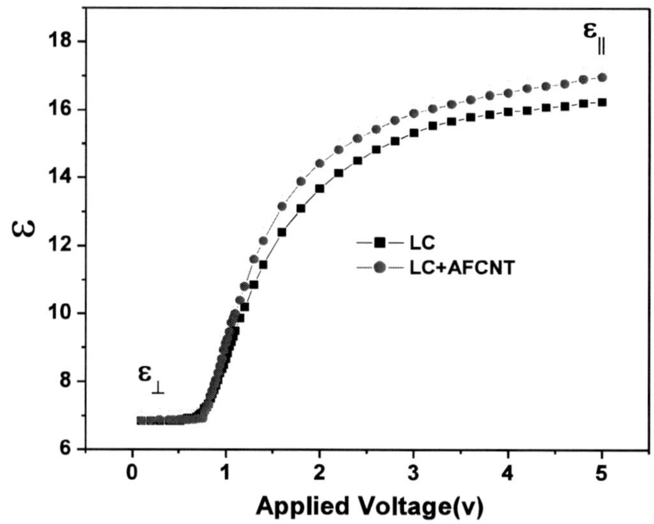

Figure 7. Voltage dependent variation of dielectric constant (ε) of pure LC (E7) and AFCNT doped LC at 2.1 kHz.

There is an enhancement in the dielectric anisotropy ($\Delta\varepsilon = \varepsilon_{\parallel} - \varepsilon_{\perp}$) of AFCNT doped LC compared to that of pure LC. The values of Δε are ~ 9.4, and ~ 10.1, for pure and AFCNT doped LC respectively. Russell et al [51] have shown that CNTs can induce alignment on the nematic liquid crystal (NLC) director field along their long axes due to LC-CNT anchoring effect. So, the enhanced Δε observed in doped LC can be attributed to the LC-AFCNT anchoring at the alignment layer. The conductivity (σ) of AFCNT doped LC is enhanced considerably compared to that of pure LC at all applied voltages as shown in Figure 8. The conductivity of the LC and AFCNT doped LC as a function of frequency at a voltage of 0.2 V is shown in Figure 9. The value of conductivity in the low frequency regime (<10 Hz) of AFCNT doped LC is less compared to that of pure LC. This indicates that the effect of ions in AFCNT doped LC cells is less compared to that of pure LC. For frequencies above 100 Hz, the conductivity of AFCNT doped LC is enhanced by a factor of 3 compared to that of pure LC. The imaginary part of the ε ($\varepsilon^{\parallel}$) is calculated using the conductivity data [48]. The frequency variation of imaginary part of $\varepsilon_{\perp}$ ($\varepsilon_{\perp}^{\parallel}$) at an applied voltage of 0.2 V is shown in Figure 10. The relaxation frequencies of pure LC, and AFCNT doped LC are found to be 3 and 40 Hz, respectively.

The dielectric spectra of liquids in the frequency range 10^{-2} to 10 Hz are determined by the near electrode processes such as the charge exchange between the electrode and charge carriers present in the LC [52]. Whereas the dielectric spectra in the middle frequency region (10- 10^5 Hz) characterize bulk properties of the sample.

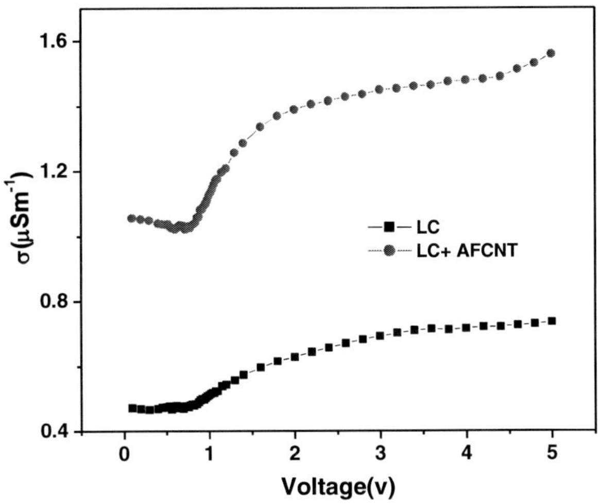

Figure 8. Voltage dependent variation of conductivity (σ) of pure LC and AFCNT doped LC at 2.1 kHz.

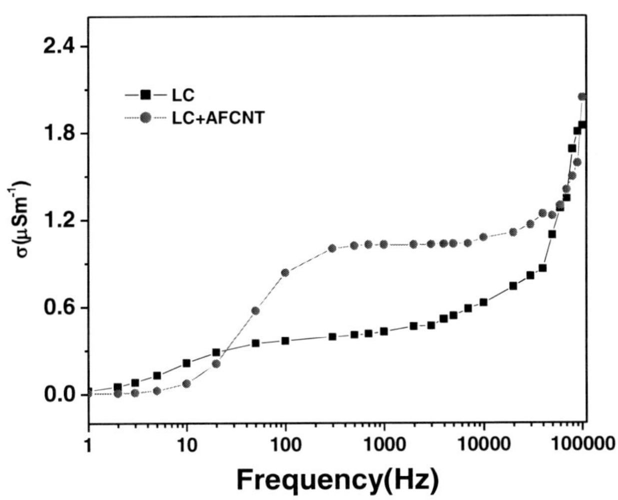

Figure 9. Conductivity (σ) of pure LC and AFCNT doped LC as a function of frequency.

Interestingly, the relaxation frequency of AFCNT doped LC were found to be 40Hz. This indicates that the increase in conductivity of AFCNT doped LC in the region above 10 Hz is not due to effect of ions, but due to the presence of AFCNT at the alignment layer. Using a dielectric spectroscopy measurement carried out on single-walled and multi-walled CNTs, Basu et al. [53] have reported that CNTs have a relaxation frequency in the range of 16 to 80Hz. They report that it is related to space charge polarization, which may involve a mechanism of charge built up at electrode-sample interface due to the presence of CNTs at the electrode. Hence, the observed relaxation frequency of 40 Hz for AFCNT doped LC can be attributed to the presence of AFCNT at the alignment layer. A strong interaction associated with the AFCNT alignment mechanism is mainly due to surface anchoring. The increase in anchoring energy is believed to be from the $\pi - \pi$ electron stacking between CNT, surface alignment layer and LC molecules [54, 55]. The AFCNT presents in LC matrix follow the orientation of the surrounding LC molecules which enhances the orientational order in the medium. The dielectric anisotropy is a measure of order parameter of the medium. Thus, the enhancement in $\Delta\varepsilon$ of AFCNT doped LC indicates an enhancement in the nematic order. The expression for threshold voltage is, $V_{th} = \pi\sqrt{K_{11}/\varepsilon_0\Delta\varepsilon}$ where K_{11} is the splay elastic constant and ε_0 is dielectric permittivity of free space [13,55]. As V_{th} of both pure LC and AFCNT doped LC are similar, the elastic constant K_{11} of doped LC is enhanced compared to that of pure LC.

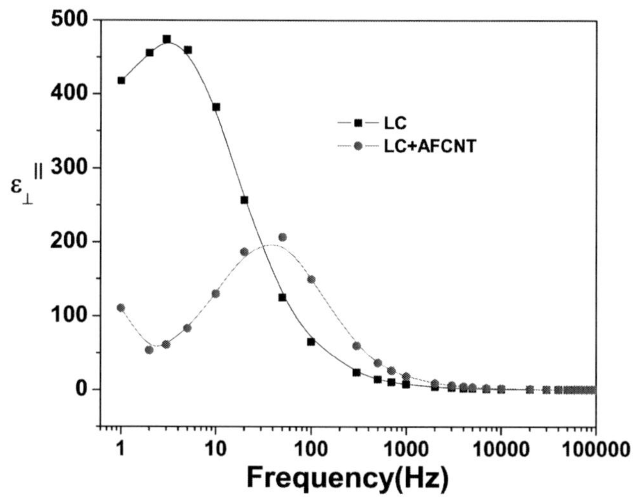

Figure 10. Imaginary part of dielectric constant conductivity ($\varepsilon_{\perp}^{\parallel}$) of pure LC and AFCNT doped LC as a function of frequency.

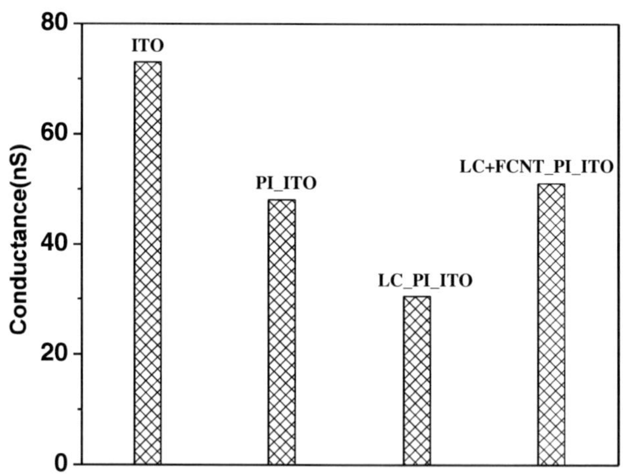

Figure 11. Average Conductivity (σ) of pure LC and AFCNT doped LC estimated using AFM in SRI mode.

In order to probe the possibility of the presence of AFCNT in alignment layer, we estimated the conductance of ITO plate, PI coated ITO, and the films of both pure as well as AFCNT doped LC deposited on the PI coated ITO substrates. The films of pure LC and AFCNT doped LC are deposited by spin coating for 30 minutes at 3000 rpm. We have measured the tunneling current (I_t) and topographic heights simultaneously in the spreading resistance imaging (SRI) [56] mode of atomic force microscope (AFM, NTMDT, Solver Pro). The bias voltage between the diamond coated conducting tips and the substrate was maintained at 0.2V.

The average value of conductance estimated using the tunneling current image profile is shown in Figure 11. The value of conductance (G) decreases with the deposition of PI on the ITO substrate. The value of G decreases further with the deposition of LC on such PI treated ITO. Interestingly, the value of G increases on depositing AFCNT doped LC as compared to that of pure LC on PI treated ITO plates. The increase in conductance in AFCNT doped LC is due to incorporation of ODFCNT in the alignment layer in the vicinity of ITO. This confirms the presence of AFCNT in the PI alignment layer.

In another study, the nature of functionalization of the SWCNT has been altered and its effect on electro-optic and dielectric measurement is investigated. In such study, the AFCNT is replaced by the ODA-SWCNT. The threshold electric field of ODA-SWCNT doped LC was found to be less as compared to that of pure LC whereas the dielectric anisotropy was enhanced. The dielectric and AFM measurements on the system confirm the presence of the SWCNTs in close proximity of alignment layer [24]. Such observation is found to be in consistent with that of AFCNTs.

In conclusion, it has been observed that incorporation of functionalized CNTs in a LC has led to improvement in the nematic ordering which is evidential from enhancement in $\Delta\varepsilon$. The observed relaxation frequency of 30-40 Hz in the functionalized CNT doped LC cell suggests the presence of SWCNTs at the alignment layer. The AFM study confirms the presence of SWCNTs at the alignment layer which can lead to better nematic anchoring which further enhances the nematic ordering.

3.2. 8CB and ODA-SWCNT System

The variation of $\Delta\mu$ of 8CB and its ODA-SWCNT nanocomposites as a function of reduced temperature (T_{NI}-T) is shown in Figure 12 (*a*). Here T_{NI} is the temperature for nematic- isotropic phase transition and T is any given temperature. The observed temperature variation of $\Delta\mu$ of pure 8CB is in good agreement with literature [57]. The value of $\Delta\mu$ is enhanced only by ~1% at a reduced temperature of T_{NI}-T = 2 °C in the highest concentration of ODA-SWCNT nanocomposite of 8CB as compared to that of pure 8CB. The value of $\Delta\mu$ enhances as the smectic phase is approached and is enhanced by ~6% at T_{NI} -T = 8 °C for the nanocomposite of 8CB as compared to that of pure 8CB. The value of $\Delta\mu$ of a liquid crystal is dependent on the π- electron conjugation of the molecular core, thus the more linearly conjugated liquid crystal would exhibit a larger optical anisotropy. In an earlier work, it was suggested that increase in the effective molecular conjugation length leads to increase in birefringence [58]. In the present work, due to the incorporation of longer ODA-SWCNTs, the effective molecular conjugation length may increase leading to increase in $\Delta\mu$.

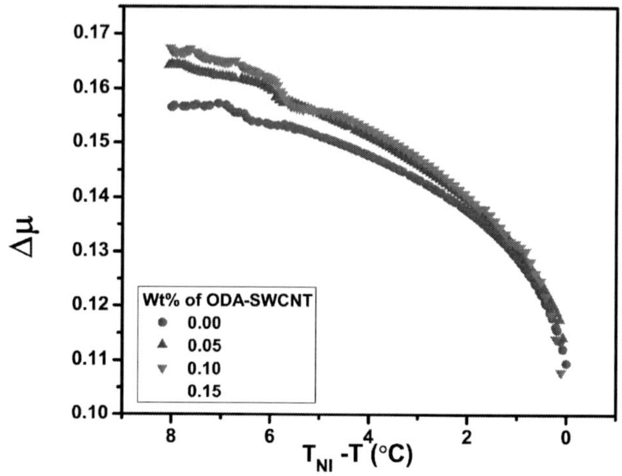

(a)

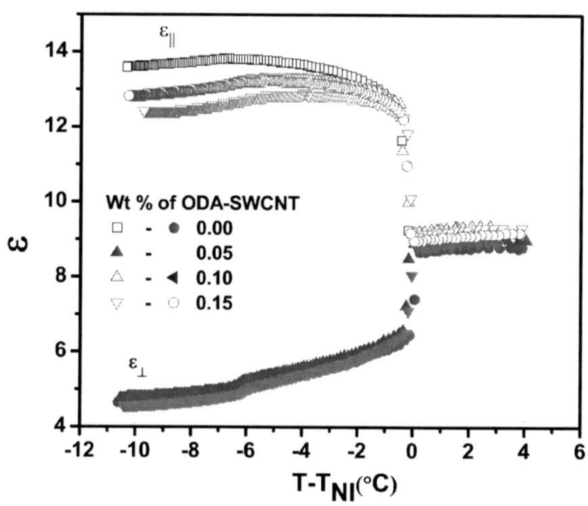

(b)

Figure 12. Variation of (*a*) birefringence (Δμ) and (b) dielectric constant (ε) as a function of reduced temperature (T_{NI} -T) for pure 8CB and its ODA-SWCNT nanocomposites. [Reprinted from DOI: 10.1080/02678292.2014.988765].

The dielectric constant for pure 8CB and its ODA-SWCNT nanocomposites as a function of reduced temperature (T_{NI} -T) is shown in Figure 12(b). The $\varepsilon_{\parallel}$ and$\varepsilon_{\perp}$ components of dielectric constant are measured using planar and homeotropic cells for pure 8CB as well as its ODA-SWCNT nanocomposites. The values of Δε for pure 8CB are in good agreement with literature [59,60]. The values of both $\varepsilon_{\parallel}$and $\varepsilon_{\perp}$decrease with increase in concentration of ODA-SWCNT. For the 0.15wt% ODA-SWCNT nanocomposite (highest concentration) at T_{NI} -T= 8 °C, the decrease in $\varepsilon_{\perp}$ is ~6% while the decrement in $\varepsilon_{\parallel}$ is ~9.5%. As a result, the dielectric anisotropy, $\Delta\varepsilon = \varepsilon_{\parallel}- \varepsilon_{\perp}$ decreases in the mesophase with increase in concentration of ODA-SWCNT (Figure 12(*b*)).

In an azoxy benzene compound [61], value of Δε decreased with decreasing temperature. This is due to the occurrence of smectic phase at lower temperatures. In this phase, the

interaction of a dipole moment with the dipoles of surrounding molecules will be influenced by the non-isotropic distribution of the centers of mass of molecules.

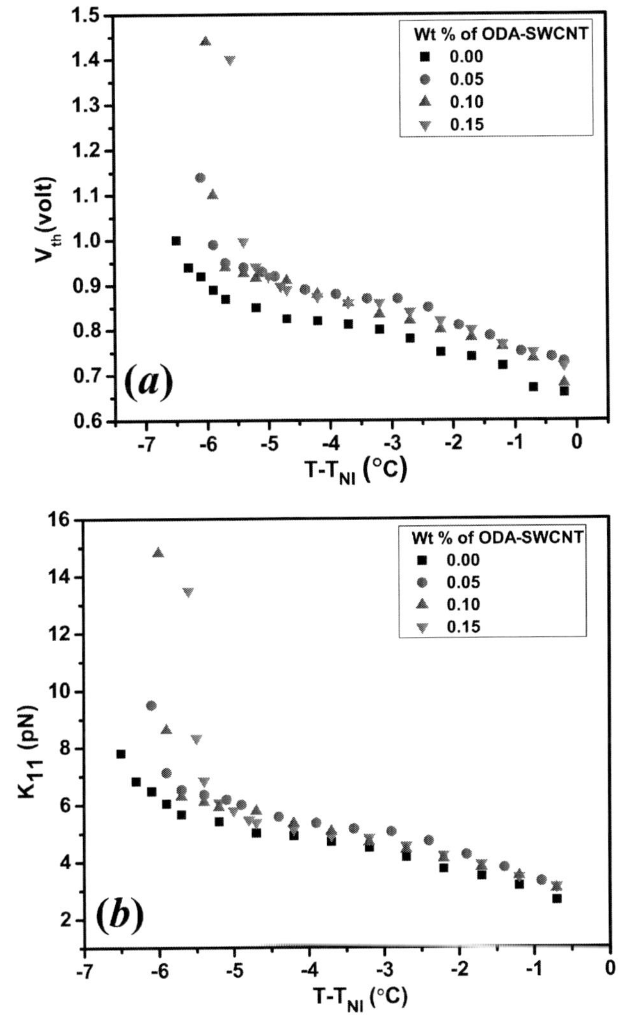

Figure 13. (*a*)Threshold voltage (V_{th}) and (b) Splay elastic constant (K_{11}) as a function of reduced temperature (T-T_{NI}) for pure 8CB and its ODA-SWCNT nanocomposites. [Reprinted from DOI: 10.1080/02678292.2014.988765].

It was suggested that as the distance between the molecules in smectic planes is much smaller than the distance between different planes, an increased anti-parallel correlation between the components of the dipole moments along the molecular axis can be expected. Consequently, the effective dipole moment in this direction is reduced leading to a decrease of $\varepsilon_{\|}$. Similarly, due to pre-smectic ordering in the nanocomposites of 8CB, the anti-parallel correlation of the dipoles parallel to the nematic director increases and as a result the effective dipole moment in the nematic phase reduces leading to decrease of $\varepsilon_{\|}$ [58, 61].

The temperature variation of V_{th} in the nematic phase for pure 8CB as well as for its nanocomposites is shown in Figure 13(*a*). The value of V_{th} at any given temperature increases with increasing concentration of the dopant ODA-SWCNT. The splay elastic constant (K_{11}) of a positive dielectric anisotropy liquid crystal in a planar aligned cell is determined using the relation, $V_{th} = \pi\left[\sqrt{K_{11}/\varepsilon_0\Delta\varepsilon}\right]$. The splay elastic constant K_{11} as a function of T-T_{NI} for pure 8CB and its ODA-SWCNT nanocomposites is shown in Figure 13(*b*). The temperature variation of K_{11} of pure 8CB compares well with that of literature [59,60]. As the concentration of ODA-SWCNT increases, the elastic constant at any given temperature increases and the effect is more pronounced on approaching towards smectic phase (i.e., T_{NI}-T =5 to 6.5 °C).

3.3. 5PCH and CdSe QDs System

The octadecyl amine capped CdSe QDs were synthesized in laboratory [39]. The size of semiconducting CdSe QDs was 3.5 nm in diameter. The CdSe QDs were dissolved in chloroform solvent and the solution was ultrasonicated for about three hours. The nanocomposite of LC, 5PCH and CdSe QDs were prepared of two different concentrations (1 and 2 wt% of QDs). The nanocomposites were used in the further studies after evaporating the solvent.

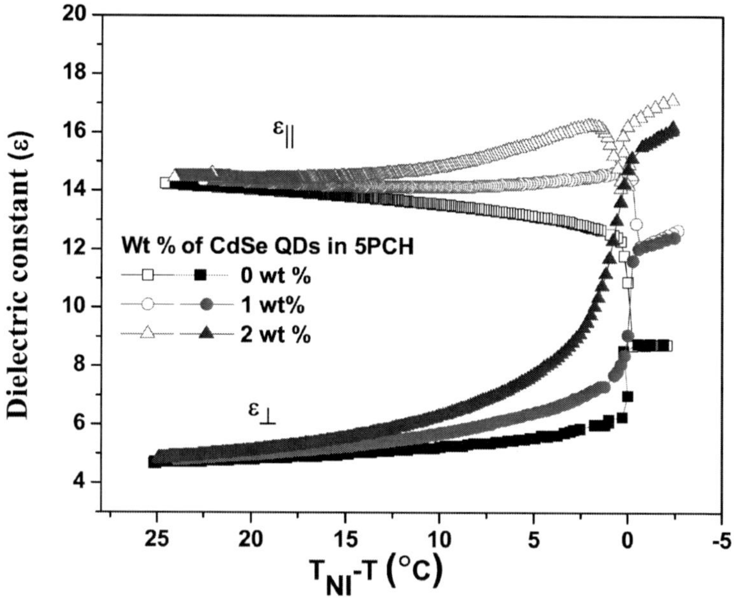

Figure 14. Dielectric constant (ε) for pure 5PCH and its CdSe QDs doped nanocomposites as a function of reduced temperature (T_{NI}-T).

On cooling pure 5PCH from isotropic to nematic phase, the dielectric constant $\varepsilon_\perp$ significantly decreases at the isotropic to nematic transition temperature (T_{NI}) and on further cooling gradually decreases with decreasing temperature. Whereas the dielectric constant $\varepsilon_\parallel$ significantly increases at T_{NI} and on further cooling gradually increases with decreasing temperature. The value of T_{NI} was determined from discontinuous change of ε value which appeared at 55.1 °C, 54.6 °C and 53.6 °C for pure 5PCH and its 1 wt% and 2 wt% CdSe doped nanocomposites, respectively. The dielectric constant for pure 5PCH and its nanocomposites doped with CdSe QDs as a function of reduced temperature (T_{NI}-T) is shown in Figure 14. The $\varepsilon_{\|}$ and $\varepsilon_{\perp}$ components of dielectric constant are measured using planar and homeotropic cells for pure 5PCH as well as its CdSe QDs doped nanocomposites. The values of both $\varepsilon_{\|}$ and $\varepsilon_{\perp}$ increase with increase in concentration of CdSe QDs. The dielectric anisotropy, $\Delta\varepsilon$ increases in the nematic phase with increase in concentration of CdSe QDs (Figure 15). The temperature variation of $\Delta\varepsilon$ of pure 5PCH is in good agreement with the values reported earlier [62,63]. The increase in $\varepsilon_{\perp}$ value indicates that the functionalized CdSe QDs with octadecyl amine group align themselves along the director of host NLC due to nematic ordering of the NLC at the surface of substrate. On application of external electric field, the dielectric permittivity $\varepsilon_{\|}$ of the medium increases due to polarizability of QDs in the system.

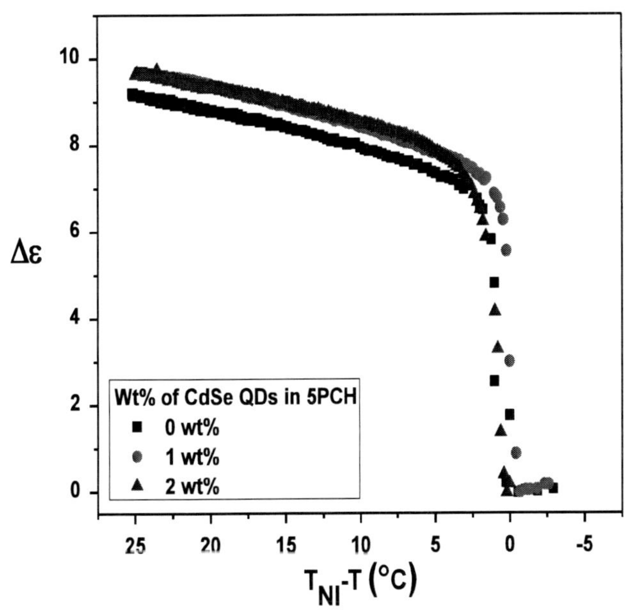

Figure 15. Dielectric anisotropy for pure 5PCH and its CdSe QDs doped nanocomposites as a function of reduced temperature (T_{NI}-T).

The temperature variation of V_{th} in the nematic phase for pure 5PCH as well as for its nanocomposites doped with CdSe QDs is shown in Figure 16. The value of V_{th} at any given temperature decreases with increasing concentration of the dopant CdSe QDs. The decrease in V_{th} may be due to the disorder created by introduction of QDs in the nematic phase which lead to formation of micro-domains [64]. The splay elastic constant K_{11} as a function of T_{NI}-T for pure 5PCH as well as for its nanocomposites doped with CdSe QDs is shown in Figure 17.

At high temperatures, V_{th} for 2wt% of CdSe QDs doped 5PCH is reduced significantly compared to that of pure 5PCH, leading to a reduction in the K_{11} of the nanocomposite as compared to that of pure LC. At low temperatures, V_{th} of QD doped 5PCH is reduced compared to that of pure 5PCH. However, the increase in $\Delta\varepsilon$ of QD doped 5PCH as compared to pure LC leads to a value of K_{11} which is comparable to that of pure 5PCH at low temperatures.

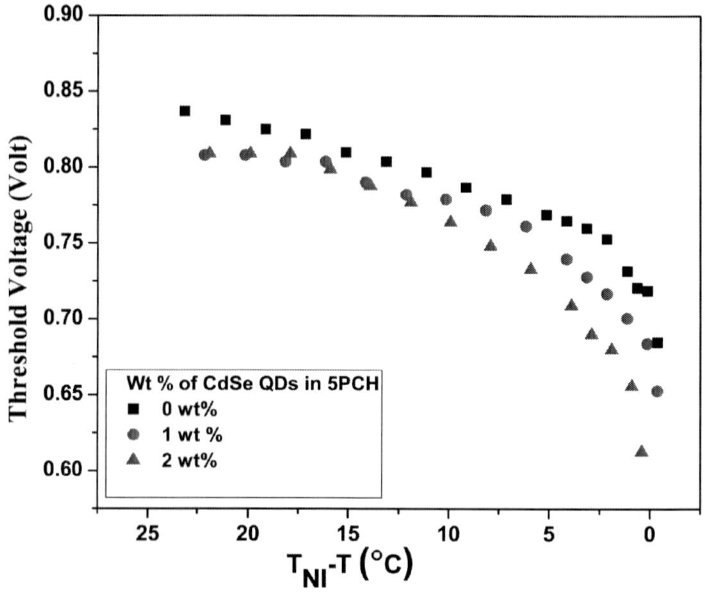

Figure 16. Threshold voltage for pure 5PCH and its CdSe QDs doped nanocomposites as a function of reduced temperature (T_{NI}-T).

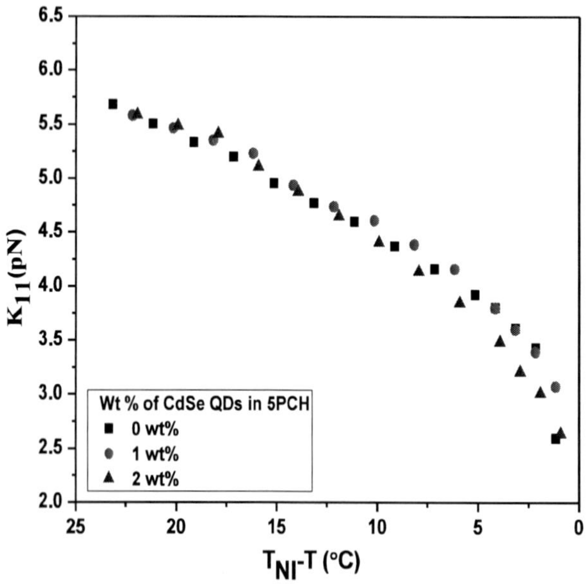

Figure 17. Splay Elastic constant for pure 5PCH and its CdSe QDs doped nanocomposites as a function of reduced temperature (T_{NI}-T).

It has been observed that above the concentration 2 wt%, the CdSe QDs start to aggregate and alignment of 5PCH NLC medium deteriorates. The aggregation of nanoparticles might be useful for some device application that requires a specific number of quantum dots. This situation may be preferred, if the aggregates of QDs are of nanoscale and anisotropic.

SUMMARY

In this chapter, we have discussed the effects of doping liquid crystals with some nanomaterials on the electro-optic and dielectric properties of the nanocomposites. We have discussed the results obtained from the systems of a few liquid crystals doped with functionalized single-walled carbon nanotubes and CdSe QDs. Due to the shape compatibility, the single-walled carbon nanotubes can mix with rod like cyanobiphenyls liquid crystal molecules. However, it was observed that pure and pristine SWCNTs do not disperse in the LC matrix. In order to improve the miscibility, the SWCNTs are functionalized with organic ligands like octadecylamine and carboxylic acid. Here, we present the change in properties of liquid crystals due to incorporation of such functionalized SWCNTs. It has been found that SWCNTs with different functionalization impose similar effects on the liquid crystal.

The SWCNTs improve the ordering in the NLC and tend to align themselves at the vicinity of the alignment layer. The role of different functionalization is to improve the degree of miscibility of CNTs in the LC matrix. The CdSe QDs doped system is found to have lower threshold voltage and an enhancement in dielectric anisotropy as compared to that of pure NLC. These results indicate that incorporation of nanomaterials into NLC provides a way to improve the performance of LC based devices.

ACKNOWLEDGMENTS

We are thankful to Department of Science and Technology, India for financial support. Thanks are also due to University Grants Commission for its support through Special Assistance Programme.

REFERENCES

[1] J. Mirzaei, M. Reznikov and T. Hegmann, *J. Mat. Chem.* 22, 22350-22365 (2012).
[2] T. Dellinger, P. Braun, *Chem. Mater.* 16 (11) 2201-2207 (2004).
[3] S. Saliba, P. Davidson, M. I. Clerc, C. Mingotaud, L. K. Myrtil and J. D. Marty, *J. Mater. Chem.* 21 (45) 18191-18194 (2011).
[4] L. Wang, X. Chen, Z. Sun and Y. Chai, *Can. J. Chem. Eng.* 85 (5) 598-601 (2007).
[5] D. F. Gardner , J. S. Evans and I. Smalyukh. *Mol. Cryst. Liq. Cryst.* 545, 3-21(2011).
[6] S. Zhang, C. I. Pelligra, G. Keskar, P.W. Mejowski, F. Ren, L.D. Pfefrelle and C.O. Osuji, *ACS Nano,* 5 (10), 8357-8364 (2011).
[7] H. Yang, M. Yang, Y. Zhang, G. Chen, *Colloid J.* 66 (5) 635-641 (2004).
[8] C. Khoo, Y. Z. Williams, B. Lewis and T. Mallouk, *Mol. Cryst. Liq.* Cryst. 446, 233-244 (2006).

[9] Y. G. Fuh, C. Y. Huang, C. K. Liu, Y.D. Chen and K.T. Cheng, *Opt. Expr.* 19 (12), 11825-11831 (2011).
[10] Y. G. Fuh, C. Y. Lin, M. S. Li and H. C. Lin, *J. Phys. D: Appl. Phys.* 45, 445104 (2012).
[11] V. G. Chigrinov, Liquid crystal devices: *Physics and Applications*, (Artech−House, Boston−London 1999).
[12] Bahadur, Liquid Crystals Applications and Uses (World Scientific, Singapore, 1990).
[13] S. Chandrasekhar, Liquid Crystals (Cambridge University Press, 1992).
[14] P. G. de Gennes, J. Prost, The Physics of Liquid Crystals (Clarendon, Oxford, UK, 1993).
[15] W. Lee, C. Y. Wang and Y.C. Shih, *Appl. Phys. Lett.* 85, 513-515 (2004).
[16] V. Manjuladevi, J. Kumar, and R.K. Gupta, (2013)., *Review in Adv. Science & Engineering Rev. Adv. Sci. Eng.* Vol 2 (4), pg - 305-312 (2013).
[17] M. Schadt , W. Helfrich, *Appl. Phys. Lett.* 18, 127- 128(1971).
[18] R. H. Chen, Liquid Crystal Displays Fundamental Physics and Technology,(Wiley Publisher, (2011).
[19] S. Naemura, Proceedings of the Third International Meeting on Information Display, Daegu, Korea, 277 (2003).
[20] Lorenz, N. Zimmermann, S. Kumar, D.R. Evans, G. Cook, and H.S. Kitzerow, *Phys. Rev. E* 86, 051704 (2012).
[21] W.T. Chen, P. S. Chen and C. Y. Chao, Jap. *J. Appl. Phys.* 48, 015006 (2009).
[22] Y.S. Ha, H. J. Kim, H.G. Park and D. S. Seo, *optics express*, 20 (12), 6448-6455 (2012).
[23] S. J. Hwang, S. C. Jeng, C. Y. Wang, C.W. Kuo and C.C. Liao, *J. Phys. D: Appl.Phys.* 42, 025102 (2009).
[24] V. Manjuladevi, R. K. Gupta and S. Kumar, J. Mol. Liq. 171, 60-63 (2012).
[25] J. Kumar, R. K. Gupta, and V. Manjuladevi, *AIP Conference Proceedings* 1536, 693-694 (2013).
[26] N. Podoliak, D. Bartzak, O. Buchnev, A. G. Kanaras and M. Kaczmarek, *J. Phys. Chem.* C 116, 12934-12939 (2012).
[27] R. Basu, G. S. Iannacchione, Phys. Rev. E: Stat., Nonlinear, *Soft Matter Phys.*, 80, 010701 (2009).
[28] H. Qi, T. Hegmann, *Liq. Cryst. Today* 20 (4), 102-114 (2011).
[29] J. Prakash, A. Choudhary, A. Kumar, D. S. Mehta and A. M. Biradar, *Appl. Phys. Lett.* 93, 112904 (2008).
[30] S.K. Prasad, K.L. Sandhya, G.G. Nair, U. Hiremath, C.V. Yelamaggad and S. Sampath, *Liq. Cryst.* 33 (10), 1121–1125 (2006).
[31] P. K. Mandal, A.Lapanik, R.Wipf, B. Stuehn, and W. Haase, *Appl. Phys. Lett.* 100, 073112 (2012).
[32] K. G. Mishra, S. chandra, and J. Gupta, *Adv. Appl. Sci. Res.* 2 (5), 212-220 (2011).
[33] S. Umadevi, X Feng, T. Hegmann, 431, 164 -175 (2012).
[34] S. Kundu, J. P. Hill, G. J. Richards, K. Ariga, A. H. Khan, U. Thupakula and S. Acharya, *Appl. Mat. & Interfac.* 2 (10), 2759-2766 (2010).
[35] S. Avinash, V. Lakshminarayanan, S. Kumar and J. K. Vij, *Chem. Commun.* 49, 978-980 (2013).

[36] P. Malik, A. Chaudhary, R. Mehra and K. K. Mehra and K. K. Raina, *Adv. Cond. Mat. Phys.* 2012, 853160 (2012).

[37] P. Kopcansky, N. Tomasovicova, M. Koneracka, V. Zavisova, M. Timko, A. Dzarova, A. Sprincova, N. Eber, K.F. Scorba T. T. Katona, A Vajda, and J. Jadzyn, *Phys. Rev. E* 78, 011702-1-5 (2008).

[38] Zhao , W. Zhou , X. Cui , Y. Tian , L. Guo , H. Yang,. *Adv. Mat* 23, 5779-5784 (2011).

[39] S. Kumar, and L. K. Sagar, *Chem. Commun.* 47, 12182-12184 (2011).

[40] V. Manjuladevi, Experimental studies on phase diagrams of liquid crystals, Ph.D Thesis (JNU, Delhi) (2004).

[41] G. Basappa and N. V. Madhusudana, *Mol. Cryst. Liq. Cryst.* 288, 161-174 (1996).

[42] Dierking I, G. Scalia G, P. Morales, *J. Appl. Phys.* 97, 044309 (2005).

[43] H. Qi, T. Hegmann, *J. Mater Chem.* 18, 3288-3294 (2008).

[44] R. Dhar, A.S. Pandey, M.B. Pandey, S. Kumar, R. Dabrowski, *Appl. Phys Express.* 1, 12501 (2008).

[45] R. Basu, *Appl. Phys. Lett.* 103, 241906, (2013).

[46] Y. J. Lim, S.S. Bhattacharya, W. Tie, H.R. Park, Y.H. Lee, S.H. Lee, *Liq. Cryst.* 40, 1202-1208 (2013).

[47] H. Lee, S. Yang, J.H. Lee, Y.S. Park, *Appl. Phys. Lett.* 104, 191601 (2014).

[48] V. Manjuladevi, N. V. Madhusudana, *Current Science* 85, 1056-1061 (2003).

[49] J. Li, G. Biard, Y. H. Lin, H. Ren, S. T. Wu, *J. SID* 13, 1017-1026 (2005).

[50] H. Ono, M. Ito, *Optics Comm.* 223, 399-409 (2003).

[51] J. M. Russell, S. Oh, I. LaRue, O. Zhou, E.T. Samulski, Thin Solid *Films* 509, 53-57 (2006).

[52] Kovalchuck, L. Dolgov, O. Yaroshchuk, Semiconductor Physics, *Quantum Electronics and optoelectronics* 11, 337-341 (2008).

[53] R. Basu, G.S. Iannacchione, *Appl. Phys. Lett.* 92, 052906 (2008).

[54] K. A. Park, S. M. Lee, S. H. Lee, Y. H. Lee, *J. Phys. Chem.* C 111, 1620-1624 (2007).

[55] S. Y. Lu, L. C Chien, *Optic. Expr.* 16, 12777-12785 (2008).

[56] P. Eyben, M. Xu, N. Duhayon, T. Clarysse, S. Callewaert, W. Vandervorst, J. Vac. *Sci. Technol. B.* 20 (2004) 471-478.

[57] P.P. Karat, N.V. Madhusudana, Mol. Cryst. *Liq. Cryst.* 36, 51-64 (1976).

[58] D.V. Sai, P. Satyanarayana, V.S.S. Sastry, J. Herman, P. Kula, R. Dabrowski, S. Dhara, *Liq. Cryst.* 41, 591-596 (2014).

[59] S. Dasgupta, P. Chattopadhyay, S.K. Roy, *Phys. Rev. E.* 63, 041703(2001).

[60] S.W. Morris, P. Muhoray, D. A. Balzarini, *Mol. Cryst. Liq. Cryst.* 139, 263-280 (1986).

[61] W.H. De Jeu, T.W. Lathouwers, P. Bordewijk, *Phys. Rev. Lett.* 32, 40-43 (1974).

[62] S. Sen, K. Kali, S.K. Roy, S.B. Roy, *Mol. Cryst. Liq. Cryst.* 126, 269 (1985).

[63] L. Pohl, R. Eidenschink, G. Krause, D. Erdmann, *Phys. Lett. A,* 60, 421 (1977).

[64] B. Kinkead, T. Hegmann, *J. Mater. Chem.*, 20, 448 (2010).

In: New Developments in Liquid Crystals Research
Editors: A. T. Petkoska and A. T. Broach
ISBN: 978-1-63482-456-9

Chapter 5

IRON(III)-CONTAINING PARAMAGNETIC LIQUID CRYSTALS: EPR AND MÖSSBAUER SPECTROSCOPY CHARACTERIZATION

Natalia Domracheva*
Zavoisky Kazan Physical-Technical Institute,
Russian Academy of Science, Russia

ABSTRACT

In this chapter the structural and magnetic properties of four different types of iron-containing liquid crystals (LC), metallomesogens, were briefly considered. EPR and Mössbauer spectroscopy were used for observing the magnetic behavior and characterization of iron-containing liquid crystals. The first section of the chapter is devoted to iron(III) complexes created on the basis of the LC poly(propylene imine) dendrimers from the first to fifth generation, which exhibit a columnar packing of the molecules in the mesophase. It was showed that iron ions existing in a high-spin state are coordinated in dendrimer ligands by two kinds of iron-complexing sites with an octahedral and a tetrahedral symmetry. Octahedral (high-symmetry) centers are located at the border of the dendrimer, whereas the tetrahedral (rhombic distorted) centers are distributed throughout all branching of the dendrimer. The octahedral "bordering" centers are exchange-coupled and are likely to form chains. The second section of the chapter deals with γ-Fe_2O_3 nanoparticles (NPs) fabricated inside LC poly(propylene imine) dendrimeric matrix, which self-organizes into a hexagonal columnar phase. It has been found that a temperature-driven transition from superparamagnetic to ferrimagnetic state was detected by EPR, wherein the blocking temperature was about 60 K. The magnetic moment and effective magnetic anisotropy of NPs have been determined from the analyses of the EPR data. The effective magnetic anisotropy constant is found to be enhanced compared to the bulk γ-Fe_2O_3 and this enhanced value is associated with the influence of the surface and shape effects. The angular dependence of the EPR signal position for field-freezing sample from liquid-crystalline phase showed that NPs possess the uniaxial anisotropy, in contrast to bulk γ-Fe_2O_3. The third section of the chapter is

* Kazan, 420029, Sibirsky trakt, 10/7, Russian Federation; Email: ndomracheva@gmail.com.

devoted to LC iron(III) Schiff base complexes with asymmetric ligands. EPR showed that complexes consist of two kinds of iron centers: $S = 1/2$ low-spin (LS) and $S = 5/2$ high-spin (HS). Interesting features were found for the monocationic Fe(III) complex, $[Fe(L)X(H_2O)_2]^+X^-$, with X = PF_6^- counterion. LS and HS centers in this compound are coupled together by antiferromagnetic exchange interactions and form a dimer structure by means of the water molecules and the PF_6^- counterion. It has been shown that this novel material has a labile LS electron configuration that switches between the $(d_{xz},d_{yz})^4(d_{xy})^1/(d_{xy})^2(d_{xz},d_{yz})^3$ ground states and is driven by the LC properties. Such interesting feature was detected in iron mesogens for the first time. The fourth section of the chapter deals with LC azomethine iron(III)complex based on the aminopyridine. It has been found that the introduction of liquid crystalline substituents in the ligand, significantly changes the spin-crossover behavior of the central magnetic Fe(III) ion and transforms all magnetic centers to the HS state. Unusual stepwise behavior of the product of the integrated EPR line intensity and the temperature with an inflection point at ~ 80 K was revealed. It has been shown that below 80 K iron(III) complexes are organized in 1D antiferromagnetic chains due to the π-π stacking of pyridine rings between the nearest neighbors with an exchange value of 2.1 cm^{-1}. Above 80 K the chains consist of dimeric molecules whose metal centers are bonded by a strong antiferromagnetic exchange interaction through oxygen bridges and the exchange value in the dimer is 162.1 cm^{-1}.

Keywords: Liquid crystals, ironmesogens, electron paramagnetic resonance (EPR), mössbauer spectroscopy

1. Introduction

The synthesis and investigation of new metal-containing liquid crystals (metallomesogens) is of great interest at present due to the possibility of combining the properties of liquid crystals (fluidity, ease of processability, one- or two-dimensional order, etc.) with the properties associated with transition metal atoms (color, paramagnetism, a electron-rich environment, etc.) [1]. Such combined systems (metallomesogens) show interesting magnetic [2,3,4] (large anisotropy, control of the molecular orientation in a magnetic field, ferromagnetism), electrical [5] (one-dimensional conductors), ferroelectric [6,7] and spin-crossover [8 - 10] properties. Liquid crystalline (LC) phases are sensitive to the external stimuli such as temperature, pressure, and electric and magnetic fields, so that the molecular orientation and LC superstructure can easily be altered [11]. In this context, there still seems to be challenging and promising perspectives in developing of inorganic soft materials, which can show a novel complexity phenomenon in response to various external stimuli by making use of LC environment. For example, control the positional order and ferromagnetism of metal-containing nanoparticles (NPs) [12] and alter the spin-variable properties of metal ions [8] due to LC properties.

Electron paramagnetic resonance (EPR) is a direct method of investigating metallomesogens [13]. EPR provides information about: - structure of the mesogenic molecules and its relationship with mesophase order; - molecular dynamics in the mesophase; - magnetic interactions between molecules and also between molecules and external magnetic fields; - structural and dynamic aspects of liquid crystal polymers. In spite of the potential advantages of this technique, the number of EPR studies on paramagnetic liquid crystals is not great. EPR spectroscopy is mostly applied to copper(II) complexes and even less frequently to oxovanadium, nickel(II), cobalt(II) and palladium(II) mesogenic materials. The variety of structures studied is also limited and includes complexes of β-diketones, salicylaldiminates and salicylaldehydes, together with some copper complexes where the metal is coordinated by N_4, O_4 or N_2O_2 donor atoms. In all of the materials mentioned above, the metal is tetracoordinated and has a planar geometry. In recent years, a rod-like metallomesogens containing 4f rare-earth metal ions, such as Pr(III), Nd(III), Tb(III), Dy(III) and Er(III) have been intensively studied [3,14,15]. Interest in these compounds arises from the fact that they possess a large magnetic anisotropy, the magnitude of which exceeds a diamagnetic anisotropy of usual liquid crystals in a hundred times. The number of EPR studies on iron(III)-containing paramagnetic liquid crystals is very limited. Domracheva *et al.* [16] reported the first example of EPR detection of magnetoelectric interactions in the LC state of the iron mesogen. The investigation of the angular dependence of the EPR spectra allows authors to display the unusual orientation of the iron molecules in the external magnetic field. It has been shown that this unusual orientation is caused by the action of the internal electric field (existing in the mesophases) on magnetic properties via the magnetoelectric effect. Ovchinnikov et al. [17] presented the EPR investigation of the first Fe(III)-containing spin-crossover mesogen. It was shown that a feature of mesogenic spin-transition system was an unusual field-induced spin instability which leads to the hysteresis of the high-spin (HS) - low-spin (LS) composition. The reason for this hysteresis was the alignment of HS-LS subsystems by magnetic field in the mesophase. Recently, Jin *et al.* [4] observed room-temperature (RT) ferromagnetism generated in a diamagnetic discotic LC phase of tetra(2-decyltetradecyloxy) phthalocyanine derivative **1** intercalated with Fe(III)-phthalocyanine [Fe(III)Pc] over the range of magnetic field of ± 2000 Oe, whereas Fe(III)Pc itself showed only a paramagnetic nature at RT. It has been shown that the origin of the RT ferromagnetism observed in weak magnetic fields was ascribed to the magnetic interactions between the conductive spins in **1** and the localized spins in Fe(III)Pc in the LC state.

In this chapter the results of new research for four different types of iron(III)-containing paramagnetic liquid crystals will be presented. The chapter is divided into four sections, according to the investigated metallomesogens, namely:

- iron(III) complexes created on the basis of LC poly(propylene imine) dendrimers;
- iron oxide (γ-Fe_2O_3) nanoparticles (NPs) formed into LC poly(propylene imine) dendrimers;
- LC iron(III) Schiff base complexes with asymmetric ligands;
- LC iron(III) azomethine complex based on aminopyridine.

Within each section a brief introduction, synthesis, characterization of the compound, identification of LC properties and the results of EPR and Möessbauer spectroscopy investigation will be given.

2. Iron-Containing Poly(Propylene Imine) Dendromesogens

Incorporation of transition metal ions under formation of dendrimeric transition-metal complexes leads to new and interesting material with specific properties, such as chromaticity, luminescence, and reversible multielectron redox processes [18]. These materials may find applications as components in molecular electronics and as photochemical molecular devices for solar energy conversion and information storage [18].

The metal centers can be incorporated in the dendrimer in threefold ways: in the core, throughout all branching, or at the periphery of the dendrimer. Example of dendrimers having metal-ion-containing core centers is given by dendritic metalloporphyrins [19,20]. The combination of a metalloporphyrin with a dendrimeric structure leads to a novel model system that mimics the encapsulation of the heme in a natural protein environment. Ferrocene-modified dendrimers attribute to the class of dendrimers with coordination centers at the periphery. These dendrimers are catalytically active [21] and useful as excellent redox sensors [22,23]. Most of the metal-ion-containing dendrimers reported so far exhibit metal complexes either in the core center or at the periphery of the dendrimer. Here, we are presenting the results of study metal-ion-containing dendrimers with the mixed way of metal incorporation.

2.1. Synthesis, Characterization of Iron Complexes and Identification of the LC Properties

The present section reports the investigation of five liquid crystalline dendrimeric Fe(III) complexes (1st to 5th generations) derived from decyloxybenzoate substituted PPI dendrimers, i.e. 3,4-bis-(decyloxybenzoyl) poly(propylene imine) by direct complexation with anhydrous iron (III) chloride [24, 25]. The general scheme of the synthesis of iron complexes with dendrimers of 1st - 5th generations, as organic ligands (L), is given below.

$$\underset{n=1-5}{\text{n-L}} + \text{xFeCl}_{3\,\text{dry}} \xrightarrow[\text{18 hours/RT}]{\text{THF}} \text{n-L-(FeCl}_3)_y + (\text{x-y})\text{FeCl}_3$$

The complexes were obtained with respect to the number of potential coordination places in dendrimeric core of ligands using five times excess of dry $FeCl_3$. The complexation reaction was carried out in tetrahydrofuran (THF) solution under argon atmosphere. For each reaction, an iron chloride solution was freshly prepared, by dissolving anhydrous $FeCl_3$ in THF and stirring it for 8 hours. A small amount of formed black precipitate was filtered off by means of a 200 nm mesh PTFE filter. The resulting clear dark-yellow solution was kept under inert atmosphere until using. The ligands were dissolved in THF and stirred 6 hours under inert atmosphere and mixed with $FeCl_3$ solution.

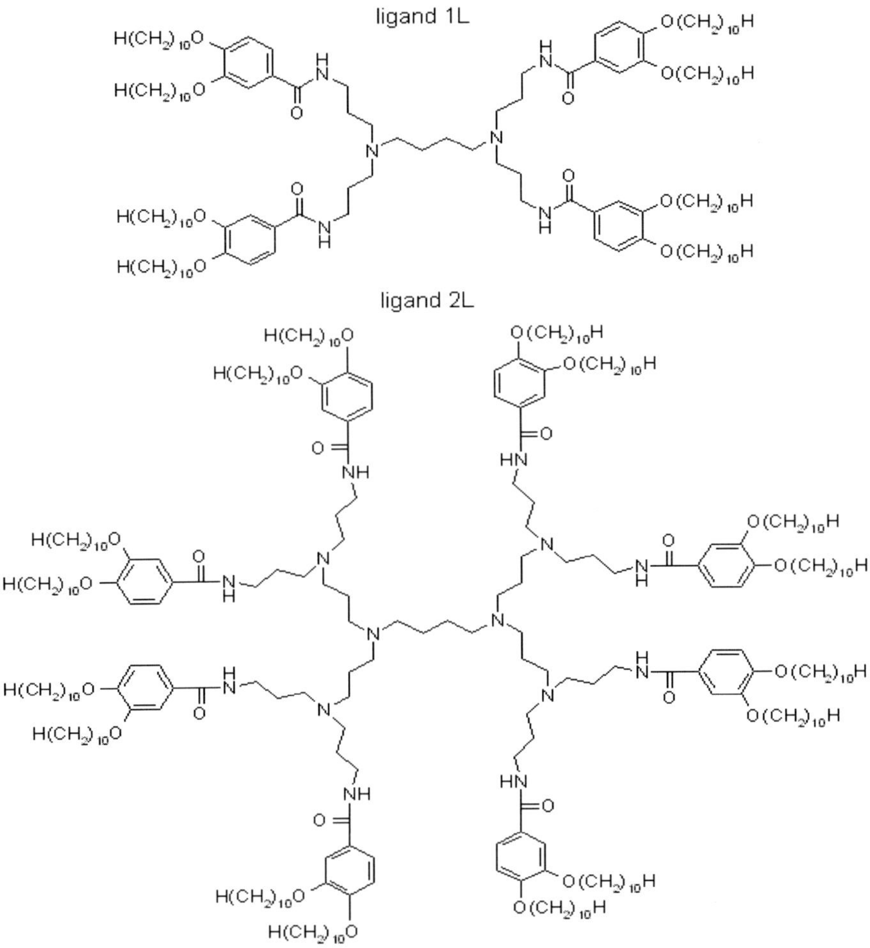

Scheme 2-1. Examples of PPI dendrimer ligands; 1st generation ligand (1L), 2nd generation ligand (2L).

After the synthesis, free $FeCl_3$ was removed by cooled ethanol, and the residue was dissolved in benzene and then filtered through teflon mesh filters (0.200 μ). The desired product was isolated by freeze drying from benzene solution. The complex results as a yellow-ochre amorphous powder.

According to the ligand generation, the dendrimer ligands are referred to as 1L, 2L, 3L, 4L, 5L (Scheme 2-1), and corresponding complexes as 1L-$(FeCl_3)_{1.8}$ (**1**), 2L-$(FeCl_3)_{4.7}$ (**2**), 3L-$(FeCl_3)_{9.5}$ (**3**), 4L-$(FeCl_3)_{21}$ (**4**) и 5L-$(FeCl_3)_{41}$ (**5**).

The purity of the investigated complexes was checked by UV spectroscopy and size exclusion chromatography (SEC, Figure 2-1) [25]. Individuality and stability was proved by IR-, Raman-spectroscopy, and mass-spectrometry [25].

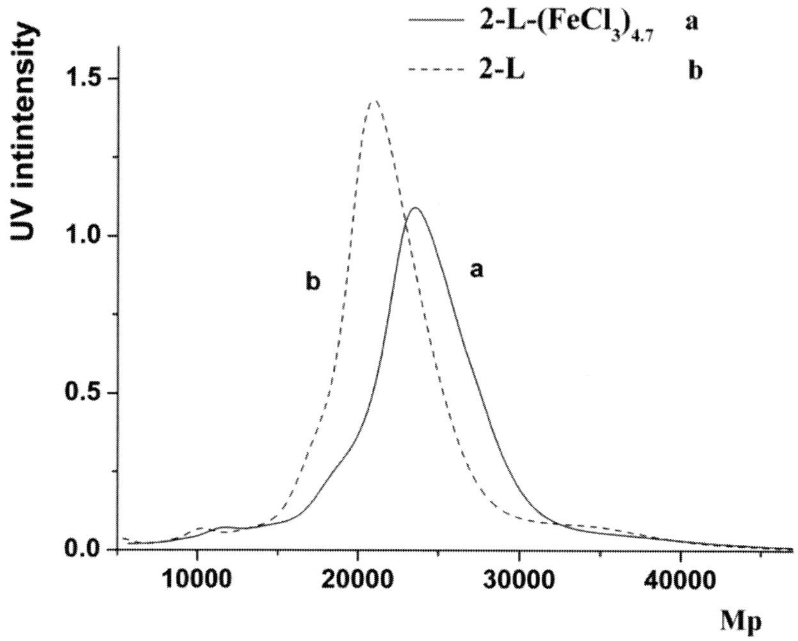

Figure 2-1. SEC chromatogram of the free 2L ligand (b) and the complex 2L-$(FeCl_3)_{4.7}$ (a).

The elemental analysis (EA) revealed the following iron content for the complexes of each generation (Table 2-1):

Table 2-1. Elemental analysis of the Fe^{III} complexes

	Compound	C%	H%	N%	O%	Fe%	Cl%
1	$C_{124}H_{216}N_6O_{12}Fe_2Cl_6$	64.48 (64.39)	9.43 (9.55)	3.64 (3.73)	8.32 (8.67)	4.84 (4.45)	9.23 (9.21)
2	$C_{256}H_{448}N_{14}O_{24}Fe_5Cl_{19}$	60.32 (61.26)	8.87 (9.65)	3.85 (3.98)	7.53 (5.78)	5.48 (5.27)	13.94 (14.06)
3	$C_{520}H_{912}N_{30}O_{48}Fe_9Cl_{33}$	61.55 (61.01)	9.70 (8.76)	4.14 (4.65)	7.51 (8.68)	5.20 (5.16)	11.84 (11.74)
4	$C_{1048}H_{1840}N_{62}O_{96}Fe_{21}Cl_{77}$	60.53 (58.80)	8.93 (9.01)	4.18 (4.78)	7.39 (9.00)	5.68 (5.51)	13.29 (12.90)
5	$C_{2104}H_{3696}N_{126}O_{192}Fe_{41}Cl_{142}$	61.28 (58.22)	9.04 (8.88)	4.28 (4.68)	7.48 (11.19)	5.66 (5.37)	12.28 (11.66)

Remark: (*) experimental data

From the EA data, the number of iron atoms $n_{EA}(Fe)$ with respect to the number of C atoms n(C) in the perfectly built dendrimeric molecule can be calculated according to:

$$\frac{n_{EA}(Fe)}{n(C)} = \frac{w_{EA}(Fe)}{M(Fe)} \cdot \frac{M(C)}{w_{EA}(C)}$$

$w_{EA}(Fe)$, $w_{EA}(C)$ are the weight fractions from EA, $M(Fe)$ and $M(C)$ are the molar masses of Fe and C atoms.

With $n(C)$ = 124 for 1L-$(FeCl_3)_{1.8}$ and 256 for 2L-$(FeCl_3)_{4.7}$, we obtained for the first generation complex $n_{EA}(Fe)$ = 1.8 and for the second generation complex $n_{EA}(Fe)$ = 4.7. The fractional number of iron atoms in 2L-$(FeCl_3)_{4.7}$ may reflect by several reason: a) a certain distribution of coordinated iron sites in the dendrimeric ligand, b) the presence of different coordination numbers and geometries in one and the same complex (as it was found for analogous copper complexes [26]) and c) the real existing defect structures in the dendrimeric ligands.

In the MALDI-TOF spectrum of 1L-$(FeCl_3)_{1.8}$, we observe - beside a peak of the free ligand - complexes with one and with two iron atoms per dendrimer. EA revealed that most of the sample should consist of the complex with two iron centres. With respect to the MALDI results, the iron content revealed by EA must be regarded as a mean value, indicating a certain distribution of the number of iron sites in the dendrimeric complex molecules. In MALDI-TOF, the whole series from one up to a four iron centres in a PPI dendrimer ligand can be observed. Peaks with higher iron content cannot be detected; they would have a too small intensity.

Results of MALDI-TOF experiments presents in Table 2-2.

Table 2-2. Data of MALDI-TOF experiment

Compound	molecular weight	
	computed value*	found value **
1L-$(FeCl_3)_{1.8}$	1983.12 [L]	1984.91 [M⁺L +1]
	2038.97 [L•Fe]	2038.71 [M⁺L•Fe]
	2061.96 [L•Na•Fe]	2061.71 [M⁺L•Na•Fe -2]
	2078.05 [M⁺L• K•Fe]	2078.13 [M⁺L• K•Fe]
	2084.95 [M⁺L•2Na•Fe]	2089.63 [M⁺L•2Na•Fe +5]
	2140.81 [M⁺L•2Na•2Fe]	2144.35 [M⁺L•2Na•2Fe +4]
	2163.79 [M⁺L•3Na•2Fe]	2167.65 [M⁺L•3Na•2Fe +4]
2L-$(FeCl_3)_{4.7}$	4106.47 [L]	4107.66 [M⁺L +1]
	4162.32 [L•Fe]	4166.04 [M⁺L•Fe +4]
	4218.17 [L•2Fe]	4219.99 [M⁺L•2Fe +1]
	4231.29 [L•3Na•1Fe]	4234.97 [M⁺L•3Na•Fe -3]
	4257.27 [L•K•2Fe]	4254.49 [M⁺L•K•2Fe]
	4274.02 [L•3Fe]	4276.53 [M⁺L•3Fe +2]
	4329.87 [L•4Fe]	4325.81 [M⁺L•4Fe -4]
	4375.85 [L•2Na•4Fe]	4381.29 [M⁺L•2Na•4Fe+6]
3L-$(FeCl_3)_{9.5}$	8353.1 [L]	8354.6 [M⁺L +1]
	8408.95 [L•Fe]	8409.68 [M⁺L•Fe -2]
	8431.94 [M⁺L•Na•Fe]	8434.47 [M⁺L•Na•Fe+3]
4L-$(FeCl_3)_{21}$	16846.59 [L]	16847.41 [M⁺L +1]
5L-$(FeCl_3)_{41}$	–	–

Remark: * ligand – [L] and amount of iron ions – [L•Fe]; ** composition of experimental molecular ion [M⁺L +n] and amount of iron ions – [M⁺L•Fe+n], where n – difference between theoretical and experimental data.

Investigation by polarizing microscopy, differential-scanning calorimetry (DSC) and X-ray scattering established that all synthesized complexes are mesomorphic [25]. The dendrimer complex of the first generation demonstrates two types of columnar packing of the molecules in the mesophase: oblique columnar (Col_{obl}) and hexagonal columnar (Col_h). All the other compounds exhibit a hexagonal columnar phase only. The temperatures of phase transitions of iron-containing dendrimers are shown in Table 2-3.

Table 2-3. Phase transition temperatures Fe^{III} complexes with substituted PPI dendrimers

Compounds	T_g [°C]	T_m [°C]	I [°C]	Phase	T_{dec} [°C]
1L-$(FeCl_3)_{1.8}$	53.0	98.5	~298.5	Col_{obl}/Col_h	~298.0
2L-$(FeCl_3)_{4.7}$	56.5	81.0	183.5	Col_h	223.0
3L-$(FeCl_3)_{9.5}$	65.5	86.5	180.5	Col_h	199.0
4L-$(FeCl_3)_{21}$	69.5	117.0	167.5	Col_h	254.5
5L-$(FeCl_3)_{41}$	78.0	125.5	189.0	Col_h	256.5

T_g: glass transition temperature; T_m: melting temperature (crystalline-mesophase transition); I: clearing temperature (mesophase-isotropic transition); T_{dec}: decomposition temperature (from termogravimetry TGA); Col_h: hexagonal columnar mesophase; Col_{obl}: oblique columnar mesophase.

2.2. EPR Characterisation of the Local Structure and Geometry of the Iron(III)-Complexing Sites in the Dendrimer Ligands

It is known that for PPI dendrimers, nitrogen atoms of the primary amine end groups and ternary amines possess strong electron-donating properties [27] and that the bis(3-aminopropyl)amine functionality acts as a strongly complexing tridentate coordinating site for the metal(III) chloride (MCl_3, with M = Co, Cr, Fe) [28, 29, 30, 31]. It has been also reported for the spherical phenylazomethine dendrimer (DPA), which imine groups act as strong coordination sites with the ferric ions, that $FeCl_3$ molecules incorporate into the dendrimer and the imine group complexes $FeCl_3$ (with equilibrium constant of complexation $K = 10^8$ [M^{-1}]), so that 3Cl^- and one imine of DPA are the ligands of Fe(III) ion [32]. Based on the literature data mentioned above we can assume that $FeCl_3$ molecules are coordinated into the substituted liquid-crystalline PPI dendrimers by two ways. Usually, Fe(III) ions form complexes with octahedral [33] or tetrahedral [32, 34] symmetry of the nearest iron surrounding. One of the aims of the present work is to identify the position and local structure of iron-complexing sites.

The electronic state of Fe(III) ions in the investigated compounds was identified by EPR and Möessbauer spectroscopy. For the complexes of different generations, EPR spectra (X-band, $h\nu_1 = 0.3$ cm^{-1}) at room temperature (Figure 2-2a) are the same and contain two main signals with effective g-factors: a more intense one with $g_1 = 2$ and a weaker one, with $g_2 = 4.3$. On decreasing temperature, an additional small line with $g_3 = 9.6$ on the left wing (Figure 2-2b) of the signal with $g_2 = 4.3$ appears.

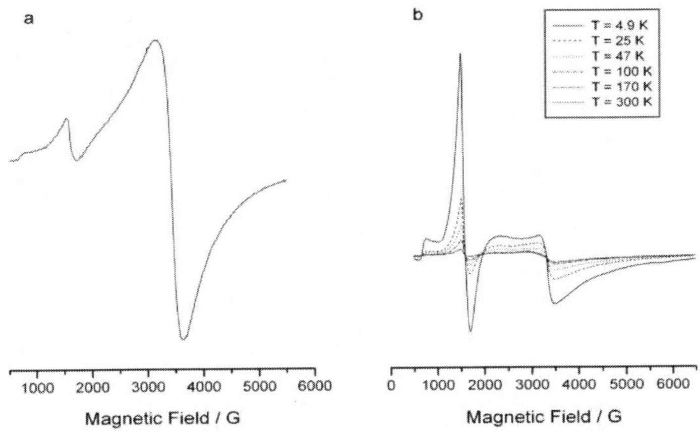

Figure 2-2. (a) X-band EPR spectra of powder complex 2L-$(FeCl_3)_{4.7}$ at 300 K and (b) at various temperatures recorded under the different amplifications.

Figure 2-3 shows temperature dependencies of the intensities and the line widths of these main signals for the Fe(III) dendrimer complexes of different generations. It can be seen that spectral parameters of the signals behave in a similar way for ligands of different generations, which testify to the same type of coordination of iron ions in the dendrimer ligands of different generations. However, the temperature dependencies of the spectral parameters of the main signals with $g_1 = 2$ and $g_2 = 4.3$ are essentially different, which demonstrate the existence of two types of Fe(III) centers in dendromesogens. As the temperature decreased, the intensity of the signal with $g_2 = 4.3$ increased much greater than the intensity of the signal with $g_1 = 2$ (Figure 2-2b, 2-3a).

It is known that such type of EPR spectrum belongs to Fe(III) ions with the ground state $^6S_{5/2}$ $(3d^5)$ and is described by the spin-Hamiltonian [35, 36]:

$$H = g\beta B_0 S + D[S_z^2 - \frac{1}{3}S(S+1)] + E(S_x^2 - S_y^2), \quad (1)$$

with $g = 2$, $S = 5/2$. The D and E are the terms of the zero-field splitting, characterising, respectively, the axial and the rhombic part of distortion of the crystal field from octahedral (or tetrahedral) symmetry; the relation $0 < E/D \leq 1/3$ holding.

The analysis of the EPR results shows that the observed lines belong to different types of paramagnetic centers. The signal with effective g - factor $g_1 = 2$ belongs to Fe(III) ions in an octahedral environment with weak ($D << g\beta B_0$) distorted crystal field, whereas the signals with $g_2 = 4.3$ and $g_3 = 9.6$ belong to iron ions with strong (orthorhombic) distorted ($D >> g\beta B_0$, $E/D = 1/3$) crystal field. The EPR spectrum of the complexes recorded in the Q ($h\nu_2 = 1.2$ cm^{-1}) band does not contain the lines with $g_2 = 4.3$ and $g_3 = 9.6$ (see inset in Figure 2-4b). This makes it possible to estimate the range of variation of the parameter of the axial crystal field: 0.3 cm^{-1} $< D <$ 0.6 cm^{-1} for the second type of paramagnetic centers [36].

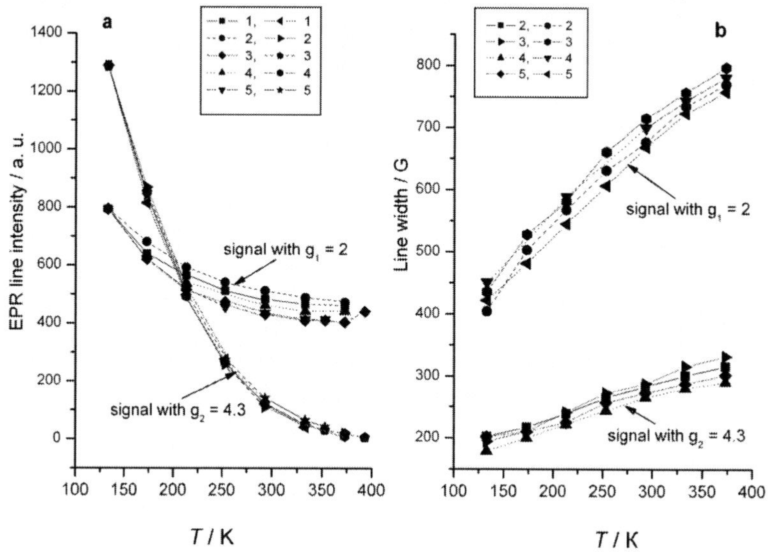

Figure 2-3. (a) Temperature dependencies of EPR line intensities (normalized to the values at 133 K) and (b) line widths for iron dendrimer complexes of the different generations.

Based on the results obtained from EPR spectra and literature data mentioned above [27, 30 - 32] the following local environment for iron ions can be suggested. The first type of paramagnetic (MA_3B_3) centers (signal g_1 = 2) is formed by coordination of three nitrogen atoms, two of which belong to amido groups, and the third one – to amino groups. The three other atoms of octahedral coordination are the chlorine atoms coming from the starting salt. This enviroment is in a good agreement with the requirements of MA_3B_3 composition of the iron(III)-complexing site, and satisfies to the elemental analysis data and mass-spectrometry. The second type of paramagnetic centers (signals g_2 = 4.3 and g_3 = 9.6) belongs to iron ions in a tetrahedral (MAB_3) environment and is formed by one tertiary amine nitrogen atom of dendrimer ligand and three chlorine atoms.

Note that such type of $FeCl_3$ incorporation into a spherical dendrimer through complexation with one inner nitrogen atom was for the first time determined in the phenylazomethine (DPA) dendrimer [32]. Iron centers of the first type are located at the border of the dendrimeric core (in the vicinity of amido end groups), whereas the second type – throughout all branching of the dendrimeric core (in the vicinity of the tertiary amino groups) (Figure 2-7).

It is interesting to note that line widths of both types of centers have practically linear dependence on the temperature (Figure 2-3b). But the line width for the bordering octahedral centers changes to a much larger extent with temperature than that one for the tetrahedral centers.

In order to check the existence of exchange interactions between iron centers, the temperature dependence of the integrated intensity of EPR line in the Q- band was investigated. The EPR intensity obtained by numerical double integration of the spectra (which proportional to the static magnetic susceptibility) follows the Curie-Weiss law $I = C/(T - \theta)$ with constant θ = - 5.7 K (Figure 2-4b). The appearance of constant θ for high-symmetrical ($D << g\beta B_0$) iron complexes of the first type (g_1 = 2) cannot be due to splitting in a zero field [37]. Estimate of the exchange integral from the value of the constant θ = - 5.7 K using the ratio $3k\theta = 2zJ\ S(S+1)$ [37] (where z is the number of nearest neighbors) leads to value J = - 0.48 K of the exchange integral (in the framework of a linear chains model). Taking into consideration the structural requirements, it is most likely that Fe(III) ions of octahedral centers (FeN_3Cl_3) interact with each other by the use of superexchange path Fe — Cl - - - Cl — Fe (involving two Cl^- anions) forming linear chains. Such structural organization of the complexes is in agreement with the structure of the columnar mesophase. Besides, it satisfies the requirements for the existence of magnetic interactions between the iron complexes and, simultaneously, the weak distortion of the octahedral field of the ion.

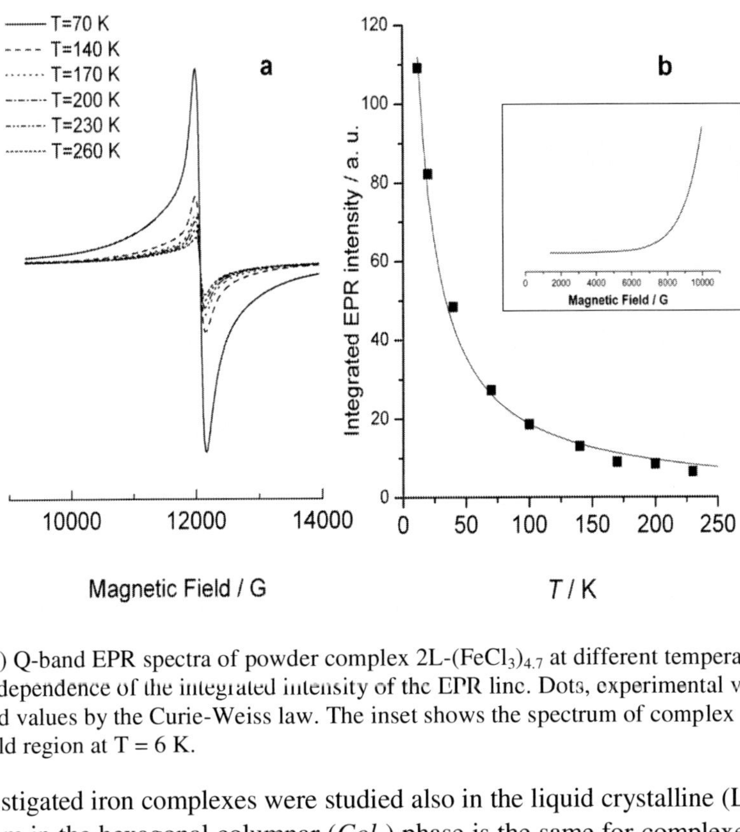

Figure 2-4. (a) Q-band EPR spectra of powder complex 2L-$(FeCl_3)_{4.7}$ at different temperatures. (b) Temperature dependence of the integrated intensity of the EPR line. Dots, experimental values, solid line, simulated values by the Curie-Weiss law. The inset shows the spectrum of complex 2L-$(FeCl_3)_{4.7}$ in the low-field region at T = 6 K.

All investigated iron complexes were studied also in the liquid crystalline (LC) state. The EPR spectrum in the hexagonal columnar (Col_h) phase is the same for complexes of different generations and contains two main signals with g_1 = 2 and g_2 = 4.3. As the temperature rises, the intensity of the signal with g_2 = 4.3 falls and isotropic phase detects only one broad line with g_1 = 2. Our attempts to resolve a fine structure in the EPR spectrum and to estimate the parameter D for octahedral centers was failed for all complexes both on heating ($T_g \rightarrow T_m \rightarrow I$) and cooling the sample from the isotropic phase in a presence of a high magnetic field (H = 0.5 T).

In order to obtain additional information on the structure of iron dendrimeric complexes, we used the orienting effect of a high magnetic field in a lyotropic columnar phase as well. Like with Cu^{II} dendrimeric complexes, [38, 39] we dissolved the 1L-$(FeCl_3)_{1.8}$ complex in a mixture of inert solvents (chloroform/toluene = 1:1), which self-organize at room temperature to a columnar mesophase. Then, this sample was sharply cooled to the glassy state in the presence of a high magnetic field ($B_0 = 0.5$ T). Figure 2-5a shows the EPR spectrum at 133 K. It should be noted, that in organic solvents without an induced lyotropic phase, the complex spectrum was similar to that of the powder state (Figure 2-2a).

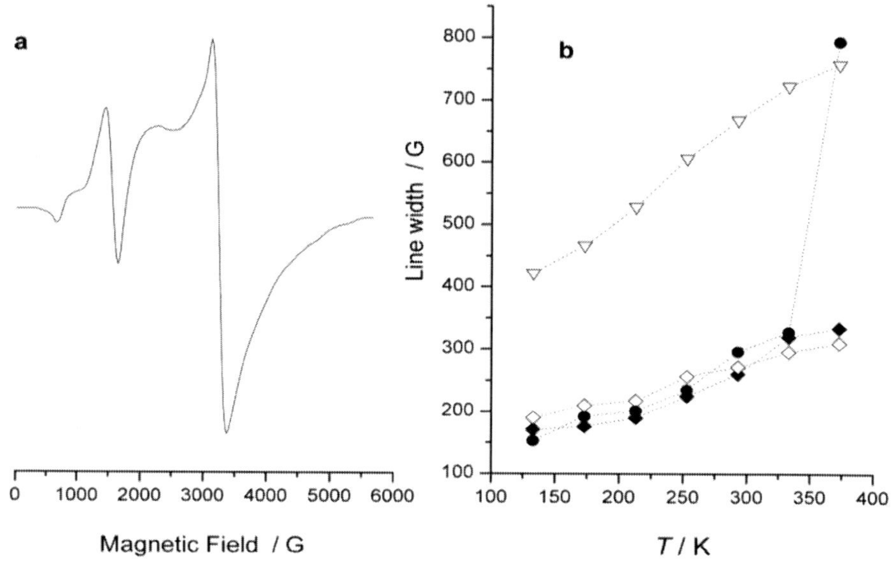

Figure 2-5. (a) EPR spectrum of complex 1L-$(FeCl_3)_{1.8}$ in the glassy lyotropic phase at T = 133 K and (b) temperature dependences of the line widths of powder complex 5L-$(FeCl_3)_{41}$ and complex 1L-$(FeCl_3)_{1.8}$ in the frozen glassy lyotropic phase. (•) – line width of $g_1 = 2$ signal for 1L-$(FeCl_3)_{1.8}$; (♦) – line width of $g_2 = 4.3$ signal for 1L-$(FeCl_3)_{1.8}$, (∇) – line width of $g_1 = 2$ signal for 5L-$(FeCl_3)_{41}$; (◊) – line width of $g_2 = 4.3$ signal for 5L-$(FeCl_3)_{41}$.

As Figure 2-5 shows, the width of the line with $g_1 = 2$ is much narrower in the lyotropic phase (almost twice) than in the powder state, while the width of the line with $g_2 = 4.3$ is practically unchanged (Figure 2-3b). Furthermore, it is interesting to note that temperature dependences of the line widths for signals with $g_1 = 2$ and $g_2 = 4.3$ are now quite similar (Figure 2-5b). However, when the lyotropic phase of complex 1L-$(FeCl_3)_{1.8}$ is destroyed (T = 373 K) and the dendrimeric complexes are randomly distributed in solution, the width of the line with $g_1 = 2$ rises abruptly (Figure 2-5b) and reaches its normal value typical for powder. The narrowing of the line with $g_1 = 2$ in the lyomesophase is attributed to the growth of exchange interactions between iron octahedral centers (FeN_3Cl_3). Low viscosity of the lyotropic phase (compared to the thermotropic phase) provides a greater flexibility of dendrimer molecules, which promotes a closer packing the dendrimer molecules in column and decrease distances between neighbouring molecules, that in turn leads to a growth the exchange interactions along the Fe — Cl - - - Cl — Fe bond.

Thus, EPR study in the lyotropic phase shows that “bordering” iron octahedral centers (FeN_3Cl_3) are exchange-coupled, whereas the “inner” tetrahedral ones ($FeNCl_3$) are isolated and not sensitive to the intermolecular magnetic interactions.

2.3. Möessbauer Spectroscopy of the Iron-Containing PPI Dendromesogen of the Fourth Generation

To confirm the EPR data complex 4L-$(FeCl_3)_{21}$ was investigated by Möessbauer spectroscopy. The gamma-resonance spectrum obtained at 80 K and 5 K (Figure 2-6) is a superposition of two quadrupole doublets. The more intensive doublet **1** (with isomer shift δ = 0.37 mm/s with respect to α-Fe and quadrupole splitting ΔE_Q = 1.20 mm/s at 80 K) has parameters typical for high spin state of Fe(III) ion in octahedral environment [40]. Doublet **2** has parameters (δ = 0.09 mm/s and ΔE_Q = 0.720 mm/s at 80 K) which is typical for tetrahedral coordination of Fe(III) irons [40].

Thus, the Möessbauer data agrees completely with EPR results and confirms additionally the existence of two kinds of Fe(III)-complexing sites with different (octahedral and tetrahedral) coordination of iron ions.

The results obtained were used to model the location of the iron-complexing centers in the dendrimer ligands from the 1st to 5th generation, fully utilizing all possible donor positions (Figure 2-7). The octahedral iron centers at the border of the dendrimeric core are marked as grey, the tetrahedral (inner) centers as black.

Examples of the location of iron-complexing centers throughout all branching of a dendrimer molecule are already known in literature for dendrimers with a phenylazomethine skeleton [41, 42]. In order to prove such an arrangement of iron centers, the Japanese scientists [42] carried out a stepwise complexation control with continuous monitoring the reaction by UV absorption spectra. This approach is very time-consuming and is not always acceptable. We show for the first time by EPR a similar location of Fe(III) centers in the dendrimer matrix of substituted liquid crystalline PPI dendrimers without being obliged to synthesize stepwise the compounds under continuous monitoring.

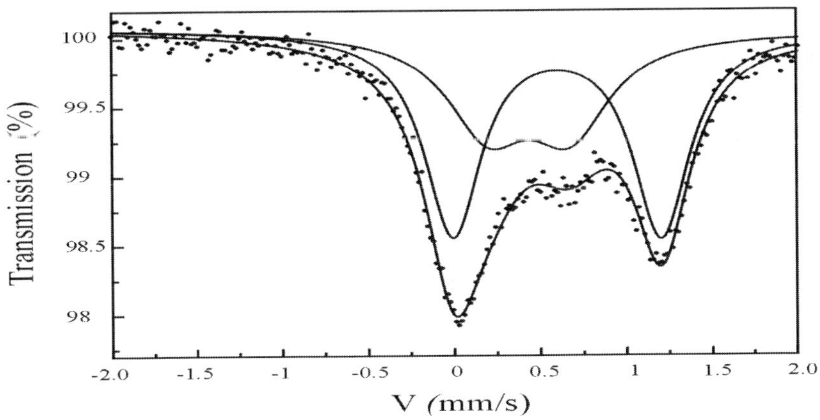

Figure 2-6. Möessbauer spectrum of 4L-$(FeCl_3)_{21}$ dendrimeric complex at 80 K.

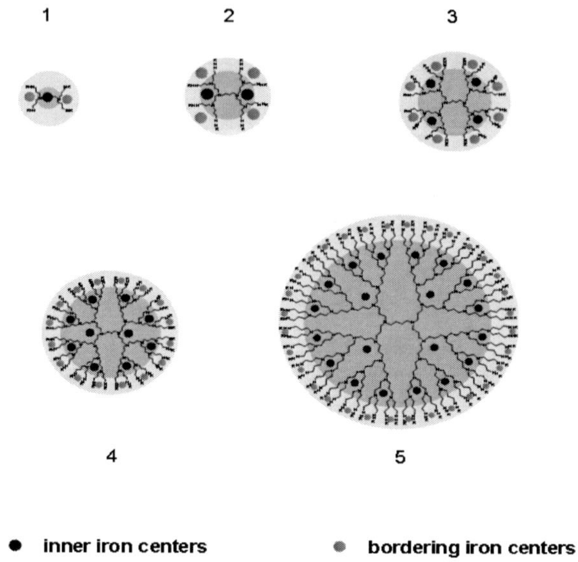

Figure 2-7. Model of location the iron paramagnetic centers in dendrimer ligands from the 1st to 5th generation.

2.4. Summary

EPR and Möessbauer spectroscopy were successfully applied for the first time to identify the local structure and geometry of Fe(III)-complexing sites in liquid crystalline bis-decyloxybenzoate substituted poly(propylene imine) dendrimers from the 1st to 5th generation [43]. It has been shown that Fe(III) ions existing in a high-spin state are coordinated in dendrimer ligands by two ways (independently of the ligand generations) and form iron centers with octahedral and tetrahedral symmetry. Octahedral (high-symmetry) centers (FeN_3Cl_3) are located at the border of the dendrimeric core (amido, amino groups), whereas the tetrahedral centers ($FeNCl_3$) with a strong rhombic distortion are distributed throughout all branching of the dendrimeric core (amino groups).

It has been found that the octahedral "bordering" centers (FeN_3Cl_3) are exchange-coupled in the lyotropic columnar phase and are likely to form chains. The "inner" tetrahedral centers ($FeNCl_3$) are isolated and do not participate in exchange interactions.

3. Iron Oxide (γ-Fe_2O_3) Nanoparticles Formed into LC Poly(Propylene Imine) Dendrimers

Magnetic nanoparticles (NPs) are key components in modern technology, with applications ranging from high-density information storage [44] to magnetic resonance imaging contrast agents [45]. Fabrication of γ-Fe_2O_3 nanoscale material with the diameter of magnetic single-domain particles ranging from 1 to 10 nm is of particular importance [46]. The size of magnetic NPs can be controlled by means of dendrimer template: unique class of monodisperse, highly branched macromolecules with highly ordered and controlled structure which is changed by dendrimer generations [47, 48]. Dendrimers successfully stabilize NPs, playing the role of nanoreactors.

For nanoscale devices not only NPs size control, but NPs self-organization is also important. Liquid crystals are a beautiful example of self-organizing nano-objects. Therefore it is very tempting to use liquid-crystalline properties to cause NPs self-organization. Substituted poly(propylene imine) (PPI) dendrimers exhibit mesomorphic properties and display the formation of columnar mesophases [49]. The aim was to use liquid-crystalline PPI dendrimers attached to iron oxide NPs as promoters to organize the positional order of NPs on the nanometer scale and to generate mesomorphism in such a combined system. Here, we are presenting the results of electron magnetic resonance (EMR) and Möessbauer spectroscopy study of a compound that we believe is the first example of ultrafine γ-Fe_2O_3 NPs fabricated inside a dendrimeric matrix, which then self-organizes into a hexagonal columnar (Col_h) phase.

3.1. Presentation of the Sample and Identification of the LC Properties

The sample obtained from the second generation poly(propylene imine) (PPI) dendrimer-iron(III) complex which has been characterized previously in [43]. The structure of the liquid crystalline PPI dendrimer of the second generation is presented in Scheme 3-1.

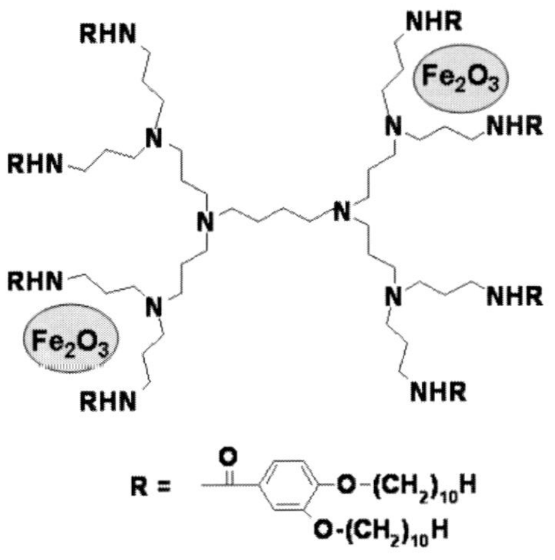

Scheme 3-1. Liquid crystalline PPI dendrimer of the second generation used to prepare the dendrimeric γ-Fe_2O_3 NPs and possible model of locating NPs in dendrimer.

The preparation of the sample is described as follows. The iron-dendrimer nanocomposite (2-D-$(Fe_2O_3)_y$) was synthesized in three stages. Following is the synthetic scheme of the nanocomposite based on the dendrimer complex derived from 3,4-bis-(decyloxybenzoyl) poly(propylene imine) (PPI) and iron (III) chloride (2-D-$(FeCl_3)_{4.7}$) of the second generation [24, 25, 43]:

$$\text{2-D-}(Fe^{3+}) \xrightarrow{N_2H_2/THF} \text{2-D-}(Fe^{2+}) \xrightarrow{NaOH/THF} \text{2-D-}Fe(OH)_2 \xrightarrow{O_2} \text{2-D-}(Fe_2O_3)_Y$$

The general procedure is set forth below. In a vacuum line 250 ml of solution of the 2-nd generation dendrimeric iron complex in THF was quickly degassed and put under Ar. Afterwards, degassed 1M N_2H_4 in THF was added and the mixture was stirred for 1.5 hours. The color of solution changed from brown to yellowish-green, which was indicative of Fe(III) to Fe(II) reduction. Addition of NaOH (in THF) to the above solution resulted in color changing from yellowish-green to olive-green, indicating the formation of $Fe(OH)_2$. The mixture was stirred under Ar for 2 hours and O_2 bubbliry for 2.5 hours. The color instantly changed to reddish-brown (immediately). The mixture was filtered through a glass filter with porosity 4. The solid residue on the filter was washed with cold ethanol until neutral pH was redissolved in 100 ml THF and filtered through a 450 mesh PTFE filter. The clear solution was evaporated to dryness. The solid residue was dissolved in 50 ml benzene, filtered through a 200 nm mesh PTFE filter, frozen with liquid N_2 and lyophilized by evaporating benzene from the still cold, solid sample under fine vacuum ($2 \cdot 10^{-2}$ mbar). The product is soft, amorphous, light-yellow compound soluble in tetrahydrofurane, benzene, chlorophorm and methylene chloride. Average yield was 55.9 %.

The purity and individuality of the desired product have been checked by electrolytic gel-permeation chromatography using UV- and IR-detectors [25].

Figure 3-1 gives three chromatograms as an example: a 2nd generation nanocomposite (2-D-(Fe_2O_3)$_y$, a complex and a ligand. Comparison of the above chromatograms makes it possible to conclude that all of the three compounds have different separations and different molecular masses, as well as different intensity registered by UV-detector [25]. Table 3-1 lists the main parameters of the chromatograms of dendrimeric nanocomposites obtained by the chromatographic analysis of the synthesized substances. The chromatography data (Figure 3-1) also support the conclusion that the final reaction of obtaining nanocomposite is completely over with a qualitative and quantitative yield.

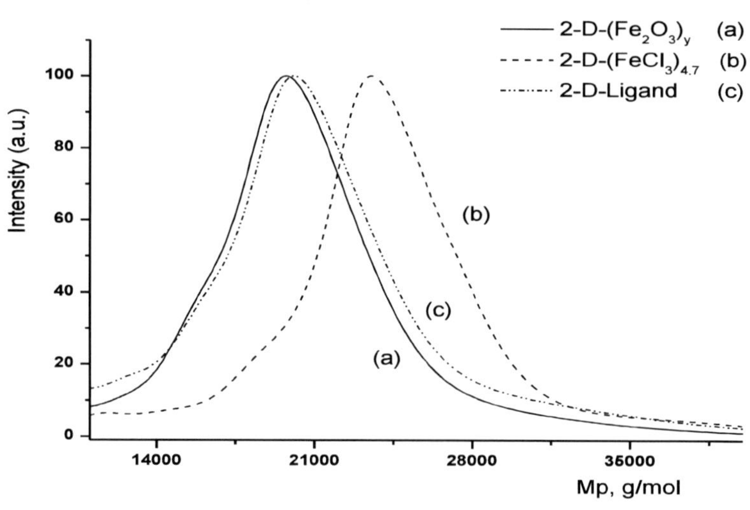

Figure 3-1. GPC- chromatograms of complex, nanocomposite and ligand.

Table 3-1. Parameters of SEC- chromatogram of dendritic nanocomposite

Compound	Mn [g/mol]	Mw [g/mol]	Mz [g/mol]	Mv [g/mol]	Vp [ml]	Mp [g/mol]
2-D-$(Fe_2O_3)_y$	$2.77\cdot10^3$	$1.59\cdot10^4$	$2.01\cdot10^4$	$1.51\cdot10^4$	19.7	$1.97\cdot10^4$

Note: Mn is the average number of the molar mass; Mw is the average mass number of the molar mass; Mz is the average molar mass; Mp is the maximum average molar mass.

Stability of the obtained iron oxide containing compound, as well as the presence of Fe_2O_3 within the dendrimeric matrix, was supported by the data of mass-spectroscopy. Table 3-2 gives the values of the main molecular ions of the nanocomposite.

Table 3-2. Mass spectrometric data for the dendronanocomposites

Compound	Molecular mass of ions	
	Theoretical	Experimental
2-D-$(Fe_2O_3)_y$	4106.47 [L]* 4266.17 [L•Fe_2O_3]* 4448.16 [L•Na•2(Fe_2O_3)]*	4105.89 [M+L -1]** 4266.11 [M+L•Fe_2O_3]** 4449.12 [M+L•Na•2(Fe_2O_3)+1]**

Note:* stands for a hypothetical composition of a molecular ligand [L] with inclusion of assumed (intended) nanoparticles [L Fe_2O_3] and ** is the composition of an experimental molecular ion [ML +n) with nanoparticles [M L Fe_2O_3+n] where *n* is the difference between the theoretical and experimental values.

Mass spectra of dendrimeric nanocomposites reveal molecular ions corresponding to the presence of iron oxide within the dendrimeric matrix. Mass spectra of a simple mixture of iron oxide and the dendrimer would exhibit a molecular ion of the dendrimer, while molecular ions corresponding to iron oxide would be found in the range of 160 to 500 units. The presence of molecular ions denrimer-iron oxide in soluble dendrimer nanocomposites is a final argument in support of the existence of their stable forms. It becomes evident that the nanocomposite is a dendrimer with particles of iron oxide between its branches.

The size of γ-Fe_2O_3 NPs was estimated by examining the transmission electron microscopy (TEM) and X-ray diffractometry data. Since the differences in electron density between the atoms of different parts of dendrimer (polyamine core, benzyl groups, alkyl shell of the PPI derivatives) are minor, no contrast in the image of pure dendrimer is observed under TEM. After obtaining nanocomposite, dark spots in the TEM image appeared. Due to high electron density of magnetic NPs, their location relative to the selected parts of dendrimer can be determined. The average diameter $\overline{D}$ of iron oxide NPs (see Figure 3-2b) estimated from TEM image is about 2.5 nm.

The hexagonal lattice constant a_h (inter-columnar distance) of the second generation dendrimeric γ-Fe_2O_3 composite was calculated from X-ray diffractometry measurements and is about 4.87 nm (see Figure 3-2a). The diameter of PPI core including the benzamide units estimated from [50, 51] is 1.86 nm. So, the length of the decyloxy chains segment surrounding the PPI core is 1.505 nm obtained by subtracting the diameter of the core of 1.86 nm from the hexagonal lattice constant a_h of 4.87 nm. According to our [43, 52] and other group results [53, 54] we may conclude that maghemite NPs are electrostaticaly coupled with the reactive (amido and amino) nitrogen periphery atoms and incorporated inside PPI dendrimer among their branches. It has been shown also that iron oxide NPs have narrow size distribution (0.5 nm) [54]. Iron oxide NPs are relatively well-separated inside the dendrimer and the suggested model NPs location is presented in Scheme 3.1.

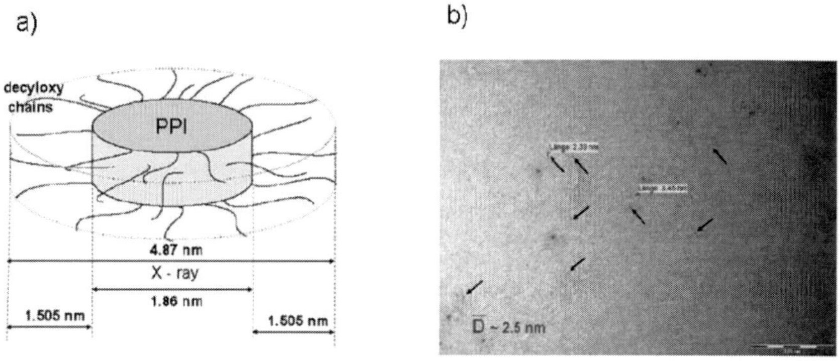

Figure 3-2. a) Schematic representation of the second generation dendrimeric γ-Fe_2O_3 NPs and dimensions of the PPI core and decyloxy chains segments estimated from X-ray diffractometry; b) TEM image of the dendrimeric γ-Fe_2O_3 NPs.

The liquid crystalline state was established by polarized microscopy, differentiating scanning calorimetry and X-ray structural analysis. The dendrimeric nanocomposite containing nanoparticles of iron oxide has mesomorphic properties. On heating in the range in which the mesophase occurs, the compound has a non-geometrical texture, Figure 3-3.

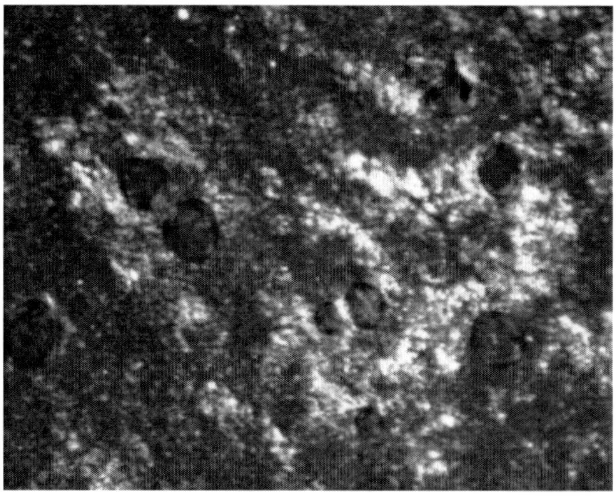

Figure 3-3. Non-geometrical texture of 2-D-$(Fe_2O_3)_y$ at heating, T = 99.5 °C.

In the cooling, the nanocomposite develops a fine (small) domain texture, Figure 3-4.

Figure 3-4. Fine(small) domain texture of 2-D-$(Fe_2O_3)_y$ at cooling , T = 95.7 °C.

According to the DSC curves of the dendrimeric nanocomposite, the mesophase occurs in the temperature range of 70 to 97 °C, the heating cycle (Table 3-3 and Figure 3-5). In the cooling cycles, the phase transition is hardly registered on the DSC curves. It may be due to a longer relaxation time of the molecules, all the more so that the mesophase in the sample on slow cooling is supported by X-ray structural analysis. Using the DSC and polarized microscopy the investigated samples of dendrimeric nanocomposites were found to vitrify on cooling.

Table 3-3. Transition temperatures of dendritic nanocomposite

Compound	*Tg* [°C]	*Tm* [°C]	*I* [°C]	Phase	*Tdec* [°C]
2-D-$(Fe_2O_3)_y$	26.0	72.5	96.5	Col_h	144.0

Note: Tg is the temperature of vitrification; Tm is the temperature of the mesophase; I – is the temperature of transition mesophase-isotrope; Tdec is the temperature of decomposition from thermogravimetric data; Col $_h$ is the columnar hexagonal phase.

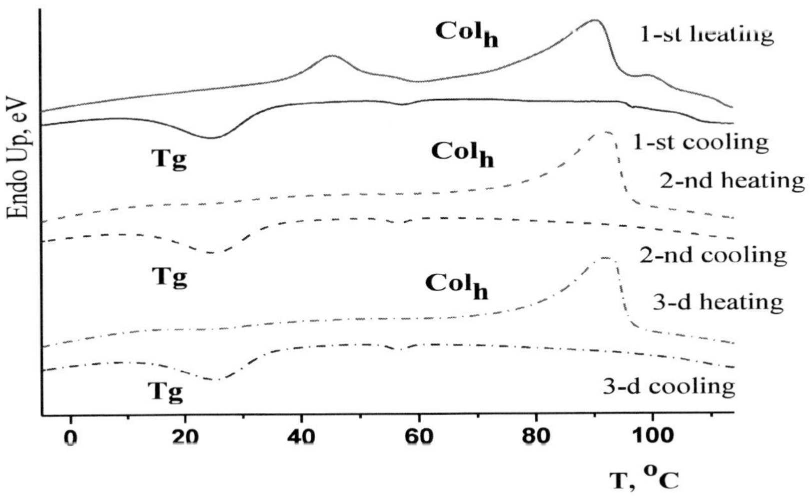

Figure 3-5. DSC-curves of 2-D-$(Fe_2O_3)_y$.

X-ray diffraction patterns of the 2nd-generation nanocomposite in the heating and cooling cycles were registered to determine the type of the mesophase (Figure 3-6 and Table 3.4).

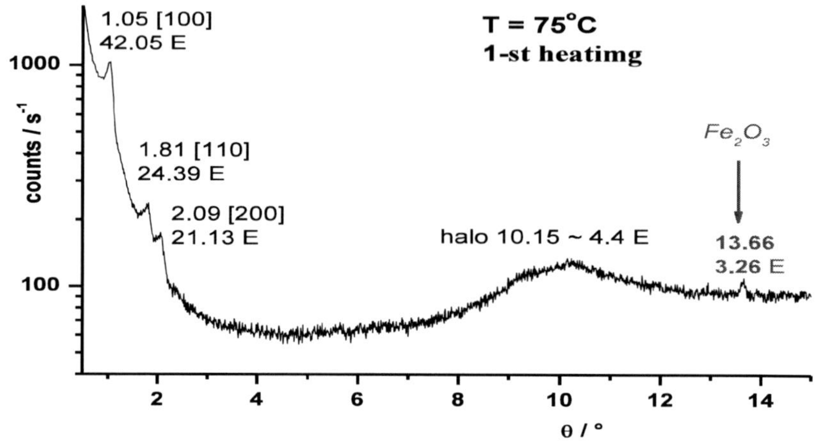

Figure 3-6. X-ray diffraction pattern of 2-D-$(Fe_2O_3)_y$ at heating.

Table 3-4. Results of calculations of X-ray diffraction pattern of 2-D-$(Fe_2O_3)_y$ at heating

Compound	T/°C	Θ/°	d_{found} / Å	$d_{calcd.}$ / Å	hkl	Phase	a/Å
2-K2.10-$(Fe_2O_3)_y$							
	75	1st heating					
		1.05	42.1	42.2	100		
		1.81	24.4	24.4	110		
		2.09	21.1	21.1	200		
		10.15	4.4	-	halo	Col_h	48.7
		13.66	3.3	Reflex γ-Fe_2O_3			
	75	2nd heating					
		1.03	42.9	42.7	100		
		1.80	24.5	24.7	110		
		2.06	21.4	21.4	200		
		10.16	4.4	-	halo	Col_h	49.4
		13.64	3.3	Reflex γ-Fe_2O_3			

The calculation of the results obtained clearly demonstrates the existence of columnar hexagonal mesophase in the iron oxide nanocomposite 2nd-generation dendrimer, derivative of poly(propylene imine).

3.2. EMR Characterisation of Iron Oxide (γ-Fe_2O_3) NPs Formed into LC Poly(Propylene Imine) Dendrimer

Electron magnetic resonance (EMR) is a very effective method of studying magnetic NPs. The theory of magnetic resonance in superparamagnetic systems was developed in [55-58]. One of the first contributions for understanding the temperature variation of the EMR spectra of NPs came from the work of Raikher and Stepanov (RS) [56]. RS developed a theoretical approach to consider superparamagnetic resonances in the limit of large Zeeman interaction as compared to the anisotropy energy. This approach is based on a rotary diffusion equation of magnetic moments motion in the presence of thermal fluctuations. In this approach, the effective field acting on NPs is the linear addition of the applied field, H, and anisotropy fields. RS results indicated that, in the superparamagnetic regime, the effective anisotropy decreased as the temperature increased, and the rate of decrease was determined by the ratio of Zeeman to thermal energies. This theory allows one to obtain physical parameters of magnetic NPs: anisotropy, magnetic moment.

Figure 3-7 shows the EMR spectra of a polycrystalline sample recorded at different temperatures from 290 to 4.2 K. The spectra are composed of a single line and the behavior is typical for that of superparamagnetic materials. One can see that upon cooling the resonance line is monotonically broadened and shifts to the lower magnetic fields. Such behavior is found in various magnetic NP systems [59-64].

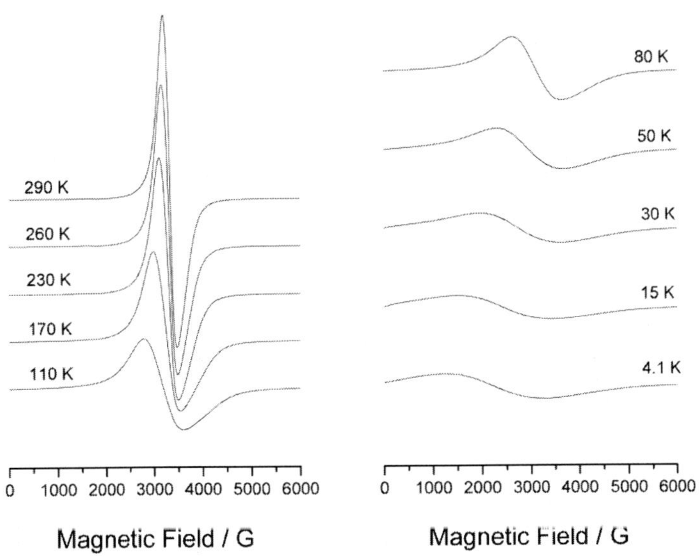

Figure 3-7. Temperature dependences of EMR spectra γ-Fe_2O_3 NPs formed into dendrimer. All spectra are recorded at the same conditions.

We additionally studied the effect of NPs alignment in the columnar phase under field-freezing (FF) conditions. NPs alignment in the mesophase was achieved by slow cooling of the sample from isotropic (375 K) to columnar (350 K) phases at an external magnetic field of $H_{fr.}$ = 7 kG. The detected alignment, which was due to the freezing field, H_{fr}, was maintained when the sample was cooled to the glassy state upon lowering the temperature. Thus, by rotating a frozen sample through the vertical axis, one can investigate the EMR line position dependence on the angle, β, between the measuring field, H_{res}, and the direction of the alignment related to $H_{fr.}$ The observed angular dependence of the line position is shown in Figure 3-8.

As the temperature decreases, the blocking effect is expected to occur for the magnetic NPs. The blocking effect has been observed and discussed for many superparamagnetic systems, but detection by EMR spectroscopy has only been reported for a few examples (see, for example, references [58, 61, 66]). The EMR line intensity obtained by numerical double integration of the experimental spectrum was measured under zero-field cooling (ZFC) condition. The sample was cooled to 4.2 K at $H = 0$ and then the EMR spectra were recorded at the heating sample up to 300 K.

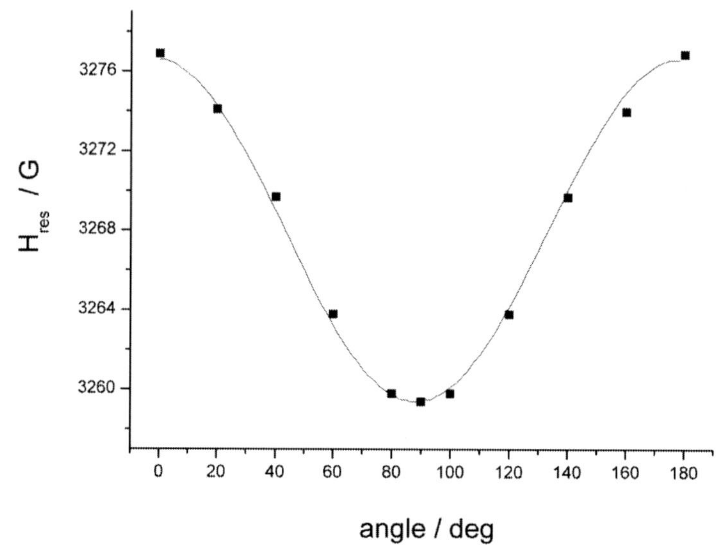

Figure 3-8. The EMR line position dependence on the angle β at T = 100 K. Full squares are experimental points and the solid line corresponds to $\cos^2 \beta$.

Figure 3-9 shows the temperature dependence of the EMR line integrated intensity for the ZFC procedure; the maximum occurs around 60 K. This dependence is similar to that observed for the ZFC susceptibility behavior in superparamagnetic systems [63, 65]. The existence of a maximum is clear evidence that the low-temperature blocking of the NPs magnetic moments takes place.

To understand the origin of the EMR line in the powdered sample, we refer to Raikher and Stepanov (RS) theory [56] for the effect of thermal fluctuation and anisotropy on the EMR in single domain NPs of dispersed ferrimagnets when the anisotropy field H_a is smaller than the external field, H. Note that the Curie temperature of bulk maghemite, $(T_c)_{bulk} = 860$ K [61], is higher than the measurement temperatures (4.2 – 380 K), and the NPs are formally in the ferrimagnetic single-domain state. When studying the magnetodynamics of single-domain particles below the Curie point, RS theory assumes that the value of the NP magnetic moment $\mu = MV$ is constant (in which V is the volume and M is the magnetization of the particle). According to the RS model, the EMR spectrum in the superparamagnetic regime strongly depends on temperature. At low temperature, the anisotropy energy barrier is larger than the thermal fluctuations ($k_BT < KV$) and so the absorption is smeared along the random distribution of the effective anisotropy fields H_a ($H_a = 2|K|/M$, here k_B is the Boltzmann constant, T is the absolute temperature, K is the effective anisotropy energy constant). On increasing the temperature, thermal fluctuations lead to a decrease of the effective anisotropy field, the temperature dependence of which is given by $H_a(T) = h_a(1/L(\xi)) - 3/\xi)$, where h_a is the "true" anisotropic field, $\xi = MVH / k_BT$, and $L(\xi)$ is the Langevin function. Because we are dealing with real systems, in which the distribution of both volumes and easy axes occurs, the absorption will be distributed along the random direction of the anisotropy axes, thus providing an asymmetric line shape at low temperature. At high temperatures, when the relaxation of magnetization through the anisotropy energy barrier is much faster than the Larmor precession, thermal fluctuations will smear out the influence of anisotropy: the resonance spectrum will be just like a normal paramagnetic spectrum with a symmetric Lorentzian line at $h\nu = g\mu_BH$.

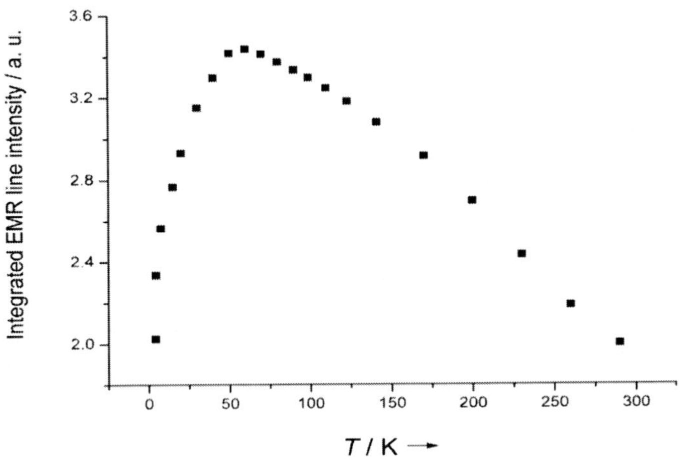

Figure 3-9. Temperature variation of the EMR line integrated intensity of γ-Fe_2O_3 NPs.

The observed EMR spectra provide an opportunity to estimate the anisotropic field (h_a), the magnetic moment of the individual nanoparticle (μ), the mean diameter of NP and the value of the anisotropy constant from the temperature dependence of the effective anisotropy field, $H_a(T)$, which was determined to be $H_{res} - H_0$, (in which H_0 = 3342 G represents the asymptotic value of H_{res} at high temperature or the value inherent to the isotropic superparamagnet). As seen from the above equation, the character of the temperature dependence of the resonance field is connected in a simply way with the nature and sign of NP magnetic anisotropy: as the temperature decreases, H_{res} increases for systems with $K > 0$ and decreases for systems with $K < 0$. The best fit of the experimental dependence (Figure 3-10) is obtained at h_a = - 1375 G and (MVH / k_B) = 77 K. Thus, the magnetic moment of the individual nanoparticle is about μ = 343 Bohr magnetons. Assuming a spherical form of the particle and taking the magnetization M = 389 emu / cm^3 [66] for the bulk γ-Fe_2O_3, we obtained values for mean volume, V = 8.18 nm^3 and mean diameter, d = 2.5 nm, for the particles. The diameter estimated coincides with the value of $\overline{D}$ found from TEM studies. From h_a, the value and the negative sign of the average anisotropy constant, $K = - 2.64 \times 10^5$ erg / cm^3, was obtained.

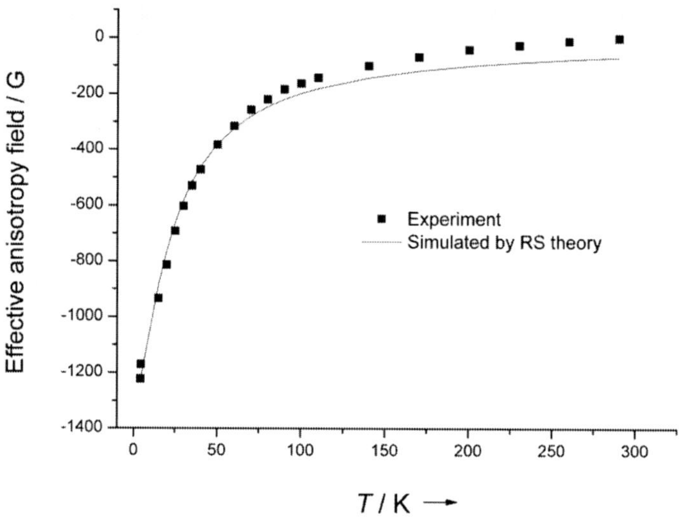

Figure 3-10. Temperature variation of the effective anisotropy field $H_a(T)$ of γ-Fe_2O_3 NPs formed into dendrimer. The solid line shows the theoretical dependence.

It is known, that the bulk anisotropy of γ-Fe_2O_3 is cubic with $K_V = -4.64 \times 10^4$ erg / cm^3 [67]. The evaluated constant, K, is an order of magnitude higher than the magnetocrystalline anisotropy (K_V). However, for noninteracting uniformly magnetized particles, the effective anisotropy constant, K, results from the combined action of magnetocrystalline anisotropy, K_V, shape anisotropy, K_{sh}, and surface anisotropy, K_S: $K = K_V + K_{sh} + K_S$. Thus, the growth of the magnetic anisotropy constant may be caused by contributions from surface and shape effects. This conclusion in favor of surface and shape contributions to anisotropy constant was obtained earlier by magnetic susceptibility measurements [68-69], as well as Mössbauer [68-69] and EMR spectroscopy [61] for small γ-Fe_2O_3 NPs embedded into other diamagnetic matrices, but not dendrimers.

Let us now consider the effect of NPs alignment in the columnar phase under field-freezing (FF) conditions (Figure 3-8). The specific angular dependence of the resonance field after the FF procedure is observed. As seen from Figure 3-8, this dependence has one minimum in the range of 0° - 180° and it fits to Equation $H_{res} = \frac{\omega}{\gamma} - H_a P_2(\cos\beta)$, in which P_2 is the second Legendre polynomial. Angular analysis allows us to conclude that NPs formed into the dendrimeric matrix possess negative uniaxial anisotropy. The anisotropy sign found coincides with the sign of magnetocrystalline anisotropy in bulk γ-Fe_2O_3 and with the sign of the "true" anisotropic field, h_a, determined from the temperature dependence of the line shift, ($H_{res} - H_0$), for the powdered sample, but in contrast to cubic anisotropy typical for bulk γ-Fe_2O_3, we observe the appearance of uniaxial anisotropy. The origin of such anisotropy may be caused by surface effects combined with shape effects (deviation of NPs from spherical shape).

Thus, RS theory showed a significant enhancement of the effective magnetic anisotropy constant dendrimeric (γ-Fe_2O_3) NPs as compared with bulk maghemite and that this enhanced value was associated with the influence of surface and shape effects. The angular dependence of the EMR signal position for the field-freezing sample from liquid-crystalline phase showed that NPs possess the uniaxial anisotropy, in contrast to bulk γ-Fe_2O_3.

3.3. Möessbauer Spectroscopy of Iron Oxide (γ-Fe_2O_3) NPs Formed into LC Poly(Propylene Imine) Dendrimer

The Mössbauer spectroscopy on ^{57}Fe was used to determine the structure heterogeneity and composition of the iron oxide NPs. The Mössbauer spectrum observed at 298 K is a superposition of a singlet and quadrupole doublet. The parameters of the single line with δ = 0 mm/s* (* indicates with respect to α-Fe at 298 K) corresponds to an α-Fe metallic core in a paramagnetic state with an area under the curve S ~ 15 % (Figure 3-11a).

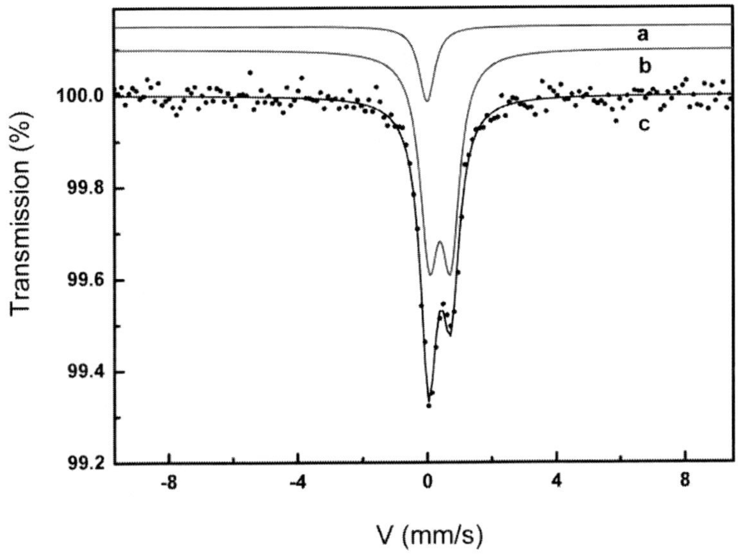

Figure 3-11. Mössbauer spectrum of core-shell iron oxide NPs at 298 K (c). The solid curves, which are located above the experimental spectrum, show the α-Fe metallic core (a) and γ-Fe_2O_3 shell (b).

The doublet with parameters (δ = 0.39 mm/s*, ΔE_Q = 0.668 mm/s, S ~ 85%) corresponds to the shell of iron oxide γ-Fe_2O_3 on the surface of NP (Figure 3-11b). The parameters of this doublet are in good agreements with the parameters for magnetic single-domain γ-Fe_2O_3 NPs encapsulated within sol-gel derived silica (SiO_2) [69].

At T = 80 K, the second doublet (Figure 3-12c) with parameters δ = 0.33 mm/s*, ΔE_Q = 3.08 mm/s, S ~ 12 % is clearly resolved. We assume that this doublet conforms to Fe(III) ions of NP bound to the dendrimer. Such Fe(III) positions are characterized by a strong temperature dependence of the Lamb-Mössbauer factor, and therefore, they do not appear in the spectrum at 298 K.

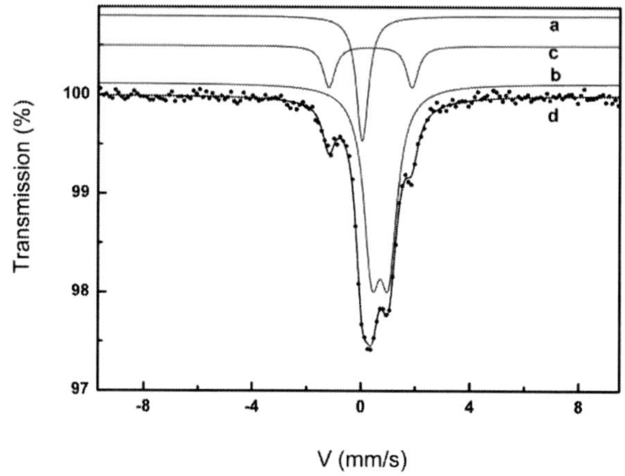

Figure 3-12. Spectrum of core-shell iron oxide NPs at 80 K (d). The solid curves, which are located above the experimental spectrum, show the α-Fe metallic core (a), γ-Fe_2O_3 shell (b) and Fe(III) ions of NP_s bound to the dendrimer(c).

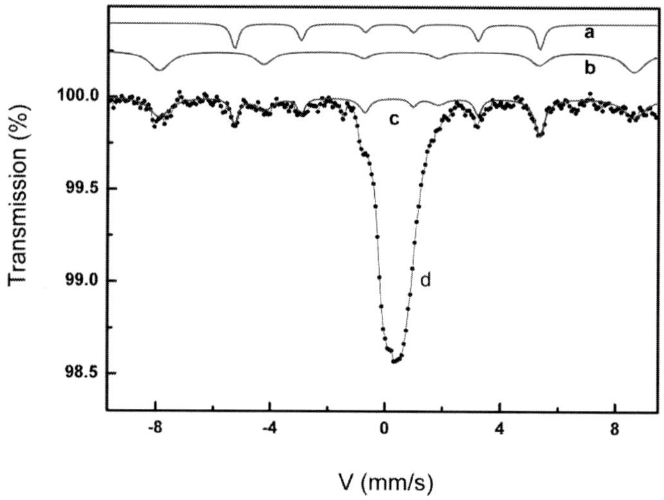

Figure 3-13. Spectrum of core-shell iron oxide NPs at 5 K (d). The solid curves, which are located above the experimental spectrum, show the α-Fe metallic core (a), γ-Fe_2O_3 shell (b), and c) is a superposition of curves a) and b).

At T = 5 K, the spectrum exhibited a magnetic hyperfine structure consisting of two superpositioned sextets. The first one (Figure 3-13a) with isomer shift of $\delta = 0.08$ mm/s* and an observed hyperfine field of 329.1 kOe, corresponds to the α-Fe core of the NPs. The observed parameters differ from the parameters of bulk α-Fe due to the defective structure of the core. The second sextet (Figure 3-13b), with parameters $\delta = 0.45$ mm/s*, $\Delta E_Q = -0.16$ mm/s, and a hyperfine field of 511.9 kOe, as we think, is responsible for the γ-Fe_2O_3 shell of the NPs [70]. A significant fraction (~ 78 %) of the superparamagnetic part relative to the total spectral absorption area (Figure 3-13d) conforms to fast dynamics in magnetic subsystems at 5 K. Such dynamics lead to transformation of the absorption lines shapes and an indefiniteness of the mathematical description of this spectrum. A considerable part of the paramagnetic (superparamagnetic) ferric ion (Fe(III)) state also confirms a weak influence of interparticle interactions on the properties of core-shell iron oxide NPs formed into dendrimers.

3.4. Summary

A second generation PPI dendrimer was used as a nanoreactor for the formation of single-domain ferrimagnetic γ-Fe_2O_3 NPs with diameter of about 2.5 nm within the liquid-crystalline dendrimer. The EMR spectra of the polycrystalline sample showed considerable broadening and shifting of the EMR line to lower fields upon cooling. This behavior was satisfactorily described by RS theory. Quantitative analysis of the temperature dependence of the effective resonance field of the EMR line found an enhanced value of the effective magnetic anisotropy constant of NPs relative to bulk γ-Fe_2O_3; this was associated with the influence of surface and shape effects [71]. The blocking temperature of the NPs magnetic moments was evaluated to be 60 K from the temperature behavior of the EMR line integrated intensity. Angular analysis performed on the field-freezing (FF) sample from the liquid-crystalline state allowed the determination of the anisotropy of the system, which was uniaxial and negative. This result led to the conclusion that this anisotropy was induced by the particle surface and shape effects [71]. Mössbauer spectroscopy determined that NPs formed were consist with a ferromagnetic α-Fe core surrounded by an iron oxide shell.

4. LC IRON(III) SCHIFF BASE COMPLEXES WITH ASSYMMETRIC LIGANDS

Six-coordinate d^5 Fe(III) complexes with Schiff base ligands are known to display a variety of magnetic behaviors. Minor variations in all of the components of the complexes can lead to high-spin (S = 5/2) or low-spin (S = 1/2) states on Fe(III) and can also lead to spin-crossover behavior [72-79]. The design, synthesis, and characterization of iron complexes with Schiff base ligands has provided useful synthetic models for bioinorganic chemistry to mimic the active sites in iron-containing proteins and enzymes [80-82]. In particular, iron(III) complexes with salicylidene amine ligands serve as a structural and electronic model for the similarly coordinated iron(III) sites found in heme-iron enzymes [83]. One of the most important characteristics of these Schiff base ligands is that even small modifications of the structure can significantly change key properties of the corresponding iron complex [80]. For example, the introduction of several electron-withdrawing nitro groups into iron complexes with salen-type ligands led to effective catalysts for hydrocarbon oxidation with dioxygen as the oxidant [84, 85]. It has been reported that the replacement of a methoxy group with an ethoxy group in a hexadentate N_2O_2 Schiff base ligand results in dramatic changes in the magnetic properties of the corresponding iron complexes due to small alterations in the intermolecular interactions and crystal packing [86].

Here, we are presenting the results of study novel iron(III) Schiff base complexes with asymmetric ligands combining properties of liquid crystals with properties associated with metal atoms (magnetism, conversion between the ground states). Liquid crystalline (LC) properties of the novel compounds were provided by rod like geometry, which was introduced by using elongated substituents in the Schiff base ligand.

A large number of papers on the six-coordinate iron(III) complexes with Schiff base ligands are known, but their liquid crystalline analogues have only been obtained comparatively recently. The first steps in the study of spin-crossover behavior for the bis(tridentate) Fe(III) complexes that exhibit liquid crystalline properties were made in ref. [8,17]. The investigation of mono(tridentate) Fe(III) Schiff base complexes with asymmetric ligands is in the early stages [87], but their liquid crystalline analogues are unknown at the present time.

4.1. Synthesis, Characterization of Iron Complexes and Identification of the LC Properties

The synthesis of three iron(III) LC complexes with the asymmetric mono(tridentate) azomethine ligand (L) and three different types of counterions, PF_6^-, NO_3^-, and Cl^- described below in accordance with Scheme 4-1, where L is azomethine 4,4'-dodecyloxybenzoyloxybenzoyl-4-salicyliden-N'-ethyl-N-ethylenediamine. The iron(III) complexes are denoted as **1**, **2**, and **3** according to their counterions PF_6^-, NO_3^-, and Cl^-, respectively.

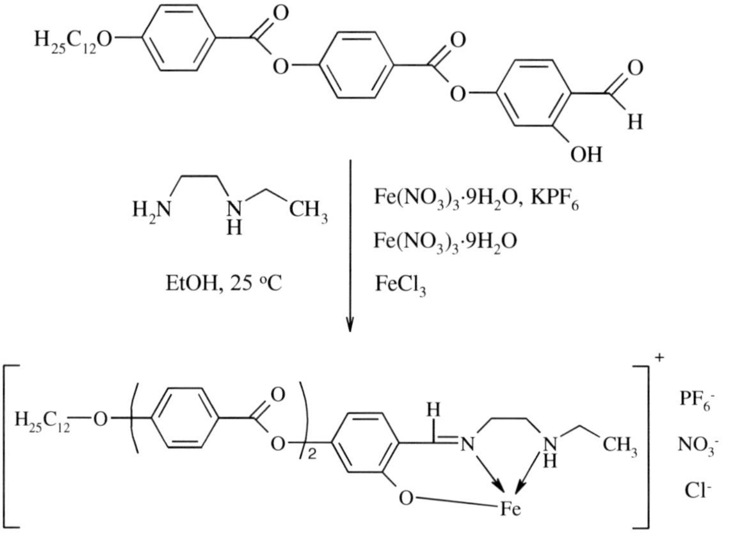

1
2
3

Scheme 4-1. Synthesis of the iron(III) LC complexes (1)-(3).

All of the synthesized compounds were characterized by thin layer chromatography, elemental analysis, IR, and NMR spectroscopy [88].

Data of elemental analysis are shown in Table 4-1.

Table 4-1. Elemental analysis of the complexes 1-3

№	Compound	M	Data of elemental analysis					
			Found, %			Calculated, %		
			C	H	N	C	H	N
1	$C_{37}H_{47}N_2O_6FeP_2F_{12}\cdot 2H_2O$	997.69	43.44	4.69	5.69	43.37	4.72	4.10
2	$C_{37}H_{47}N_5O_{15}Fe\cdot 5H_2O$	947.77	46.79	6.12	6.09	46.84	6.16	7.38
3	$C_{37}H_{47}N_2O_6FeCl_2\cdot 2H_2O$	778.68	55.05	5.59	2.88	57.04	6.21	3.59

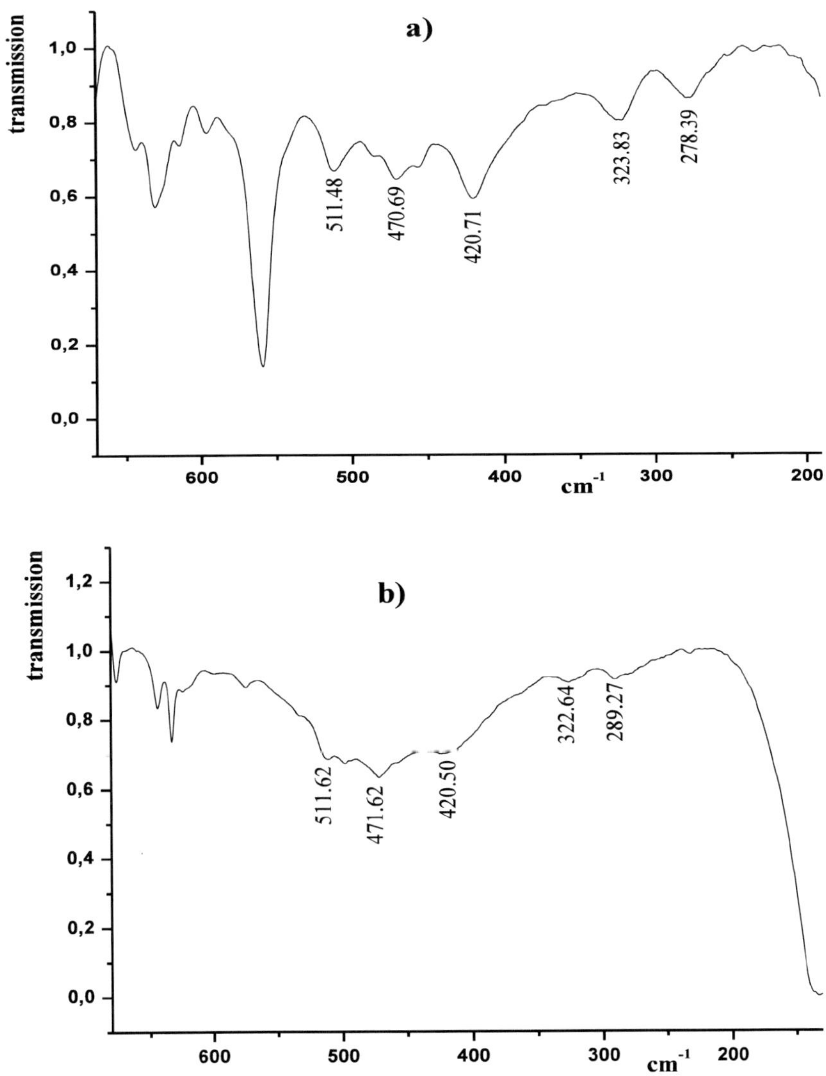

Figure 4-1. (Continued).

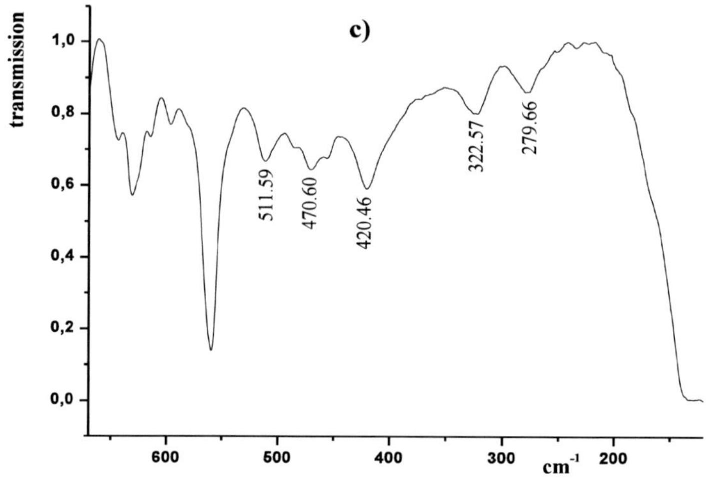

Figure 4-1. FT-IR spectra of complexes in the far region (680-140 cm^{-1}): a) $C_{37}H_{47}N_2O_6FeP_2F_{12}\cdot 2H_2O$ (1), b) $C_{37}H_{47}N_5O_{15}Fe\cdot 5H_2O$ (2), c) $C_{37}H_{47}N_2O_6FeCl_2\cdot 2H_2O$ (3).

The presence of the complex-forming ion was confirmed by FT-IR spectra in the far region [89]. The structure of the compounds, which was established by matrix-assisted laser desorption/ionization time-of-flight (MALDI ToF), is in good agreement with data obtained from elemental analysis.

Table 4-2. Data of MALDI-TOF experiment

Ion	m/z	
	Experimental	Theoretical
*$C_{37}H_{47}N_2O_6FeP_2F_{12}$ (**1**)*		
$[L^+]$	617.42	616.78
$[L\cdot Fe\cdot NH_4]^+$	687.65	690.66
$[L\cdot Fe\cdot 2NH_4\cdot K]^+$	743.82	747.69
$[L\cdot Fe\cdot PF_6\cdot K]^+$	846.77	856.59
$[L\cdot 2PF_6]^+$	907.96	906.69
$[L\cdot Fe\cdot 2PF_6\cdot H_2O]^+$	983.25	980.43
*$C_{37}H_{47}N_5O_{15}Fe$ (**2**)*		
$[L^+]$	618.73	616.78
$[L\cdot Fe\cdot NH_4]^+$	688.79	690.63
$[L\cdot Fe\cdot K]^+$	707.42	711.71
$[L\cdot Fe\cdot NO_3]^+$	738.34	734.62
$[L\cdot Fe\cdot NO_3\cdot K]^+$	777.65	773.62
$[L\cdot Fe\cdot 2NO_3\cdot Na]^+$	825.10	819.63
$[L\cdot Fe\cdot 3NO_3\cdot 3H_2O]^+$	912.46	912.62
$[L\cdot Fe\cdot 3NO_3\cdot K\cdot 5H_2O]^+$	984.14	988.62
*$C_{37}H_{47}N_2O_6FeCl_2$ (**3**)*		
$[L^+]$	617.75	616.78
$[L\cdot Fe\cdot Cl]^+$	703.61	708.08
$[L\cdot Fe\cdot 2Cl]^+$	743.73	743.53
$[L\cdot Fe\cdot 2Cl\cdot 2H_2O]^+$	785.4	779.53
$[L\cdot Fe\cdot 2Cl\cdot K\cdot 2H_2O]^+$	823.6	818.63

The molecular structure of the complexes **2** and **3** was formed by coordination of a mono(tridentate) ligand (L), a water molecules and counter anions (NO_3^- or Cl^-). Compound **1** is a monocationic mono(ligand) Fe(III) complex, $[Fe(L)X(H_2O)_2]^+X^-$, where X = PF_6^-, in which the iron(III) ion has an octahedral configuration formed by three ligator (ONN) atoms of the tridentate ligand, one counterion and two water molecules (Scheme 4-2).

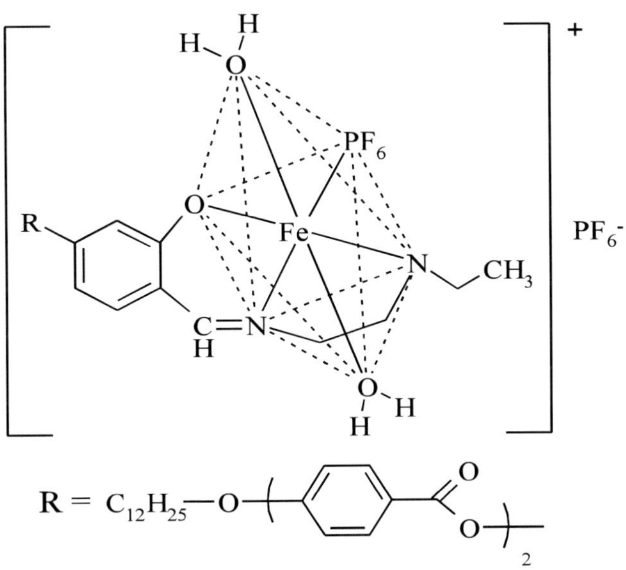

Scheme 4-2. Schematic representation of the structure for complex 1.

Analogous coordination has been found in the research of Leovac [87] for a similar non-mesogenic iron(III) complex **30** with an asymmetric Schiff base ligand and the structure of complex **1** is based on analogy to ref. [87].

The LC properties were studied by differential scanning calorimetry [88]. Phase transition temperatures and corresponding transition enthalpies of LC Fe(III) complexes are given in Table 4-3.

Table 4-3. Phase transition temperatures of complexes

№	Compound	T_{ph1} [°C]	ΔH_1 [J/g]	T_{ph2} [°C]	ΔH_2 [J/g]	T_{melt} [°C]	ΔH_{melt} [J/g]
1	$C_{37}H_{47}N_2O_6Fe$ $P_2F_{12}\cdot 2H_2O$	114.03	0.55	135.47	*	142.05	5.39*
2	$C_{37}H_{47}N_5O_{15}Fe$ $\cdot 5H_2O$	98.11	3.29	114.11	3.92	141.75	19.04
3	$C_{37}H_{47}N_2O_6Fe$ $Cl_2\cdot 2H_2O$	99.67	0.61	111.34	0.59	-	-

Remark: * total area under curve.

Table 4-4. Experimental and calculated EPR parameters for the LS iron(III) complexes

Complex	g_x	g_y	g_z	A	B	C	k	Δ/λ	V/λ	$\Delta E_{12}/\lambda$	$\Delta E_{13}/\lambda$
1 (of the first-type)	-2.215	-2.215	1.935	0.1007	0.9939	0	0.848	2.677	0.007	7.662	8.597
1 (of the second-type)	1.975	-1.98	-2.165	0.7436	0.0005	0.6657	0.793	-5.506	1.489	8.984	21.056
1 (in N-8)	-2.209	-2.228	1.945	0.089	0.9951	0.003	0.954	2.866	0.107	8.123	9.259
2	-2.22	-2.23	1.948	0.0925	0.9954	0.0017	0.935	2.817	0.062	8.042	9.049
3	-2.225	-2.23	1.935	0.0936	0.9938	0.0008	0.9646	2.7989	0.0331	8.012	8.971
$Fe(L)_2PF_6$	-2.146	-2.233	1.969	0.0715	0.9973	0.0141	0.992	3.584	0.743	8.553	13.101

4.2. EPR Characterization of LC Iron(III) Schiff Base Complexes with Asymmetric Ligands

Similar X-band EPR spectra (Figure 4-2) were observed for all of the compounds in the temperature range of 4.2 to 300 K and consisted of two types of signals: a low-field signal with $g_{eff.}$ = 4.3 that arose from high-spin (HS) Fe(III) ions and a high-field signal that corresponded to the low-spin (LS) Fe(III) species. Magnetic parameters of these signals were independent of temperature. The EPR spectrum of the first type belonged to Fe(III) ions that have the $^6S_{5/2}$ ground term and that are described by the spin Hamiltonian (1). The observation of resonance at $g_{eff.}$ = 4.3 indicated [36] that the high-spin Fe(III) ions have a strong ($D >> h\nu$) low-symmetry ($E/D \approx 1/3$) crystal field ($h\nu$ is the microwave energy and is ca. 0.3 cm^{-1} at X-band frequencies). The EPR spectrum of the second type is typical of S = 1/2 Fe(III) ions that have the $^2T_2$ ($t_{2g}{}^5$) ground term and that are described by the spin Hamiltonian of rhombic symmetry, see Equation (2), where the g values are close to 2.

The observed g tensor values are given in Table 4-4. It should be noted, that the values found are typical of those found for the low-spin iron (III) complexes [90-92].

$$\hat{H} = \beta(g_x\hat{B}_x\hat{S}_x + g_y\hat{B}_y\hat{S}_y + g_z\hat{B}_z\hat{S}_z) \quad (2)$$

The calculation of the integrated EPR intensity for the HS and LS centers showed that the ratio between these centers depends on the type of counterion (the fraction of the HS centers being minimal for Cl^- and maximal for PF_6^- counterions). However, the thermally driven spin-crossover transition from a spin state S = 1/2 to a state S = 5/2 was not detected for the studied compounds under investigation.

Compound **1** with PF_6^- counterion behaves in a different way. The EPR spectra, together with the temperature dependence of their intensity for compound **1**, presented an exceptional case among the investigated iron complexes. These features will be considered below.

We first tried to assign the magnetic (principal) g axes to the molecular symmetry axes for the LS iron complex **1**, since such information was unknown and crystals suitable for an X-ray structure determination could not be grown. For this purpose the method of orientation of paramagnetic molecules in the anisotropic solvent was used as described in ref. [38, 93]. The nematic liquid crystal (LC) N-8 (eutectic mixture: 4-methoxy- and 4-ethoxy-benzylidene-4'-*n*- butylaniline) was employed as the anisotropic solvent. The investigated paramagnetic molecules have a rod-like geometry due to the elongated alkyl chains (tails). Such rod-like molecules are known to align in nematic liquid crystals along the director ***n*** (the characteristic direction of the sample orientation) owing to the interaction with the LC matrix of the elongated alkyl chains. Note, that the magnetic g tensor axes are mainly determined by the nearest neighbors of the iron ion (the distorted octahedron).

The orientation of the LS paramagnetic molecules was observed by sharply cooling a solution of complex **1** in the N-8 matrix in the presence of a high magnetic field (B_0 = 8000 G). The observed orientation, which is characterized by a director ***n***, is due to the field B_0. This is maintained when the sample freezes to the glassy state upon lowering the temperature. Thus, one can investigate the EPR line intensity dependence on the angle ξ between ***n*** and ***B*** by rotating the sample with respect to the external magnetic field. If the molecular y axis is oriented along $\boldsymbol{n}$, then the spectral line related to this axis of the complex has to be of maximal intensity for the initial orientation ($\xi = 0^0$) and minimal intensity after rotation of the sample by 90^0. Theoretical calculations [94] and experimental data [95] have also demonstrated that the EPR line intensity must have a maximum in the angular dependence at intermediate ξ values, if the magnetic axes of the g tensor and the molecular axes do not coincide.

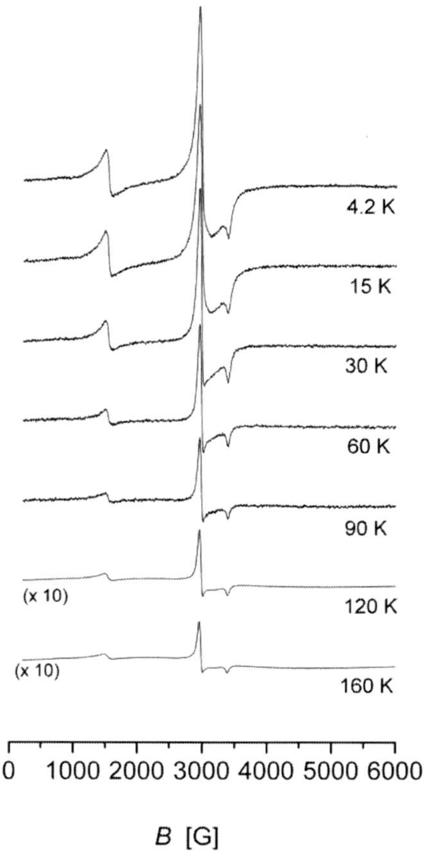

Figure 4-2. The EPR spectra of the polycrystalline complex 2 at different temperatures; ν = 9. 23 GHz.

The monotonous change of the EPR line intensities with the angle ξ variation (Figure 4-3a) clearly shows that the directions of the molecular and the principal g tensor axes coincide. As seen in Figure 4-3(a), the y component of the g tensor LS iron complexes **1** decreases, and the z component increases when the sample is rotated through 90^0. Such spectral features indicated that the molecular y axis was oriented along the director $\boldsymbol{n}$. For a qualitative analysis of the EPR spectra observed, we use the theory developed for paramagnetic probes in the oriented liquid crystals [94, 95]. The EPR spectra were simulated by using a home-written program with algorithms described in ref. [94, 95]. For nematic liquid crystals the orientational distribution function has an exponential form and is expressed as Equation (3):

$$P(\psi,\varphi) \approx \exp(-R_x \sin^2\psi \sin^2\varphi - R_y \sin^2\psi \cos^2\varphi - R_z \cos^2\psi), \quad (3)$$

where ψ, φ are Euler angles that connect the molecular coordinate system and the sample coordinate system (determined by the director $\boldsymbol{n}$) and R_i (i = x, y, z) are the effective orientational parameters.

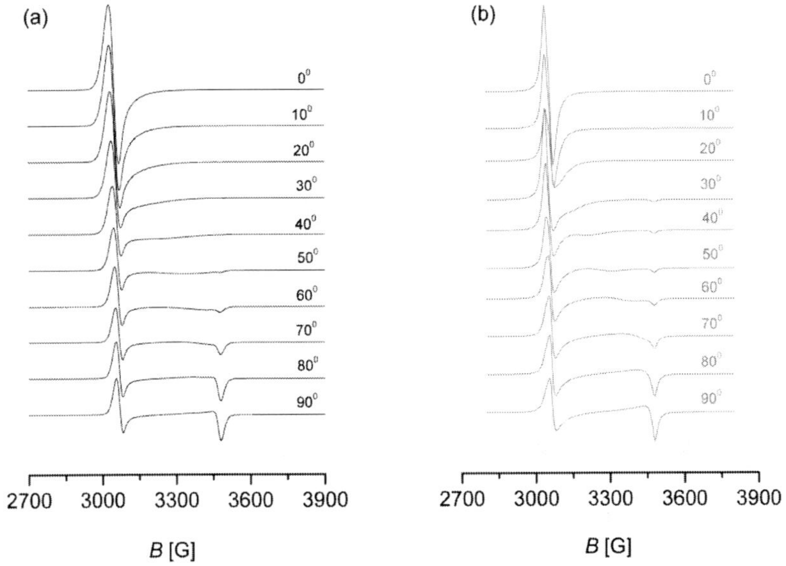

Figure 4-3. a) The experimental and b) the simulated EPR spectra for complex 1 at a low temperature in a eutectic mixture of N-8, which was previously oriented in the magnetic field (B_0 = 8000 G) for different angles ξ (a curve parameter).

Figure 4-3(b) shows the simulated EPR spectra of the LS complex **1**, calculated with the following orientational parameters R_y = 0, R_x = 6, and R_z = 7 and magnetic parameters g_x = 2.209, g_y = 2.228, g_z = 1.945. The individual Lorentzian line shape and linewidth of 15 G was used for the simulation. The degree of orientation of the molecular (i = x, y, z) axes of the LS complex **1** were S_y = 0.71, S_x = - 0.36 and S_z = - 0.35.

It is known that the mono(tridentate)-coordinated Schiff base ligand has an almost planar conformation from the structural data for a similar non-liquid-crystalline iron(III) complex **30** with an asymmetric Schiff base ligand [87]. Thus, based on the orientation results of the LS complex **1** in the nematic liquid crystal (N-8), we concluded that the long molecular y axis of the LS complex **1** lies in the (ONN) coordination plane and its direction is determined by the elongated alkyl tails, the x axis is perpendicular to y and also lies in this plane; while the molecular z axis is perpendicular to the (ONN) plane.

From the known crystal structure for the analogous non-liquid-crystalline complex **30** [87], we were able to postulate that two water molecules occupy the *trans* position to the coordination (ONN) plane in the LS complex **1**, and that one PF_6^- counterion is ionically bound to the Fe(III) ion and occupies the fourth position in this plane, whereas the other PF_6^- counterion lies in the second coordination shell.

Some interesting features of the polycrystalline complex **1** are detailed below. The EPR intensities of both the LS and the HS centers of complex **1**, which was obtained by numerical double integration, exhibit a significant temperature dependence with a maximum at $T_N = 7$ K (Figure 4-4a). Above T_N the EPR integrated intensities follow the Curie-Weiss law with a paramagnetic Curie-Weiss constant θ_1 = -14.1 K and θ_2 = -15.1 K for the LS and the HS centers (Figure 4-4b, 4-4c), respectively. The negative sign of θ indicates an antiferromagnetic coupling between Fe(III) ions. In accordance with the integrated intensity ratio, the number of LS to HS molecules is approximately 1:1. If we normalize the integrated intensity of each molecule fraction and superimpose these curves, identical magnetic behavior of the LS and the HS centers is observed (Figure 4-4a). This shows that the LS and the HS iron(III) ions are coupled together and form a dimer structure, where the PF_6^- counterion and the water molecules are, probably, the bridging units.

Another interesting feature of complex **1** is the change in the number of low-spin (n_{LS}) molecules with respect to high-spin (n_{HS}) molecules with an increase in temperature. As seen in Figure 4-5a, the ratio only changes after 260 K when it begins to increase up to 340 K and then monotonically decreases to the melting point.

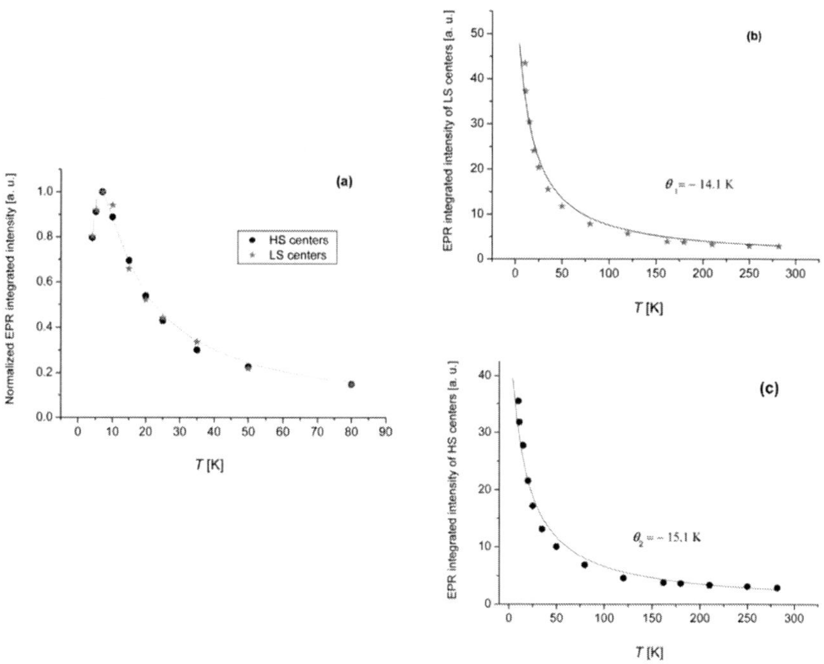

Figure 4-4. a) The temperature dependence (4.2 K ≤ T ≤ 80 K) of the EPR normalized integrated intensity of the HS and LS centers; b) the experimental (★) and the simulated (—) temperature dependence of the EPR integrated intensity of the LS centers at 9.5 K ≤ T ≤ 280 K. The solid line corresponds to the Curie-Weiss fit with θ_1 = - 14.1 K; c) the experimental (•) and the simulated (—) temperature dependence of the EPR integrated intensity of the HS centers at 9.5 K ≤ T ≤ 280 K. The solid line corresponds to the Curie-Weiss fit with θ_2 = - 15.1 K.

The temperature dependence of increasing the number of n_{LS}/n_{HS} centers can be characterized by the equilibrium constant (K) and is described by Equation (4):

$$\ln K = \ln(n_{LS} / n_{HS}) = -\frac{\Delta H}{RT} + \frac{\Delta S}{R}. \tag{4}$$

The thermodynamic parameters [enthalpy (ΔH) and entropy (ΔS)] of that process, ΔH = 0.76 kJ/mol and ΔS = 0.0056 kJ/mol, were calculated from the straight line given by plotting ln K vs. 1/T (Figure 4-5b). The obtained enthalpy value is very small and is close to the energy needed for *cis* / *trans* isomerization position of the water molecules (0.42 kJ/mol), if the energy of these molecules changes only because the geometry of the compound changes [96]. Thus, we can assume that an increase in the number of LS centers to HS centers is likely to be caused by the transition of one water molecule from the axial (*trans*) position to the (*cis*) position in the coordination (ONN) plane.

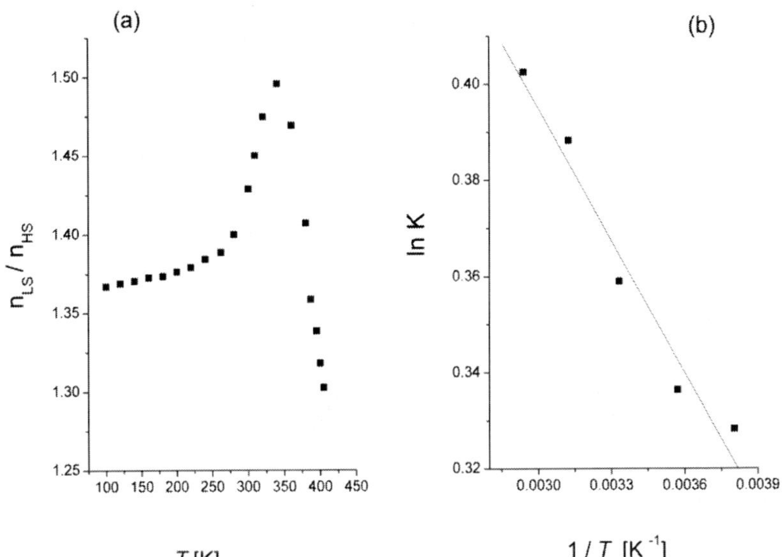

Figure 4-5. a) The temperature dependence of changing the number of LS/HS centers for complex 1; b) ln K vs. 1/T for complex 1. The solid line was calculated by means of Equation (4).

The next important feature of complex **1** is the appearance of the second-type of LS centers that is characterized by the magnetic parameters: g_1 = 1.975, g_2 = 1.98 and g_3 = 2.165. This new type of LS centers is best observed in the liquid crystalline phase at high (387 – 405 K) temperature (Figure 4-6b), but a small fraction of these centers are registered also at low temperature (Figure 4-6a). The EPR spectra simulation enabled us to separate both types of LS signals and to trace the dynamics of changing the number of the second-type LS centers (n_2) relative to the first ones (n_1) with temperature increase by using numerical double integration of the corresponding parts of the spectrum (Figure 4-7a).

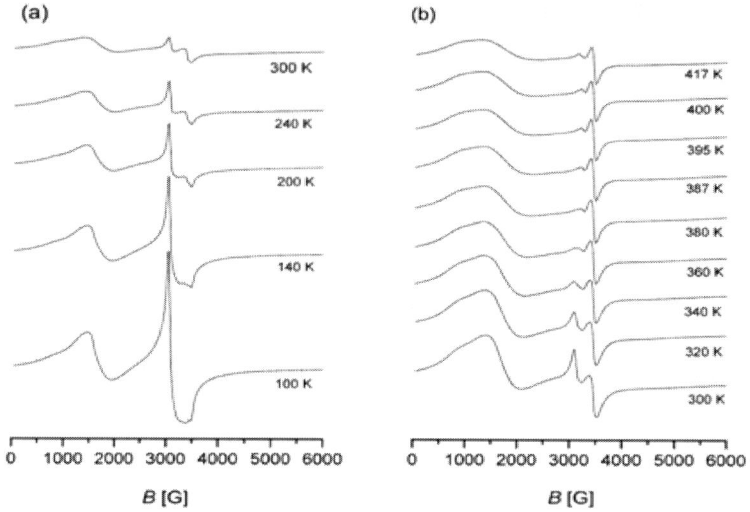

Figure 4-6. The EPR spectra of the iron(III) complex 1 at a) low and at b) high temperatures.

Before considering this dynamic effect, the determination of the ground states for the LS centers of complex **1**, **2** and **3** will be examined. The theory of the *g* tensors of LS d^5 ($t_{2g}{}^5$) systems in distorted octahedral environments, was developed by Bleaney and O'Brien [97], discussed by Griffith [98] and applied to a large number of LS complexes [38, 76, 90-94]. The d^5 system is treated as a d^1 system via the hole formalism. Under the combined action of axial (Δ), rhombic (V) ligand-field distortions, and the spin-orbit interaction (λ), the sixfold degeneracy of the $^2T_2$ ($t_{2g}{}^5$) state is converted into three well-separated Kramers' doublets. The lowest Kramers' doublet in which resonance is observed has the form of Equation (5),

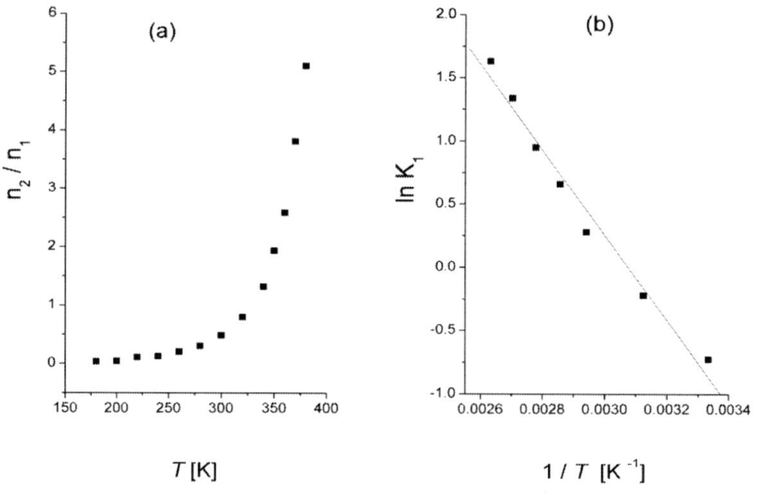

Figure 4-7. a) The temperature dependence of changing the number of the second-type of LS centers with respect to the first-type of LS centers for complex 1; b) ln K_1 vs. 1/T. The solid line was calculated by means of Equation (4).

$$\psi_1 = A\left|+1^+\right\rangle + B\left|\xi^-\right\rangle + C\left|-1^+\right\rangle,$$
$$\psi_2 = A\left|-1^-\right\rangle - B\left|\xi^+\right\rangle + C\left|+1^-\right\rangle, \tag{5}$$

where ψ_1 and ψ_2 are linear combinations of the $3d$ atomic orbitals associated with the t_{2g} level:

$$\xi = id_{xy} = (\left|+2\right\rangle - \left|-2\right\rangle)/\sqrt{2}; d_{xz} = -(\left|+1\right\rangle - \left|-1\right\rangle)/\sqrt{2};\ d_{yz} = i(\left|+1\right\rangle + \left|-1\right\rangle)/\sqrt{2}.$$

From the first-order interaction of these states with the magnetic moment operator ($k\boldsymbol{L}$ + $g_e\boldsymbol{S}$) the expressions (6) for the principal g -values are obtained:

$$\begin{aligned} g_z &= -2[A^2 - B^2 + C^2 + k(A^2 - C^2)], \\ g_x &= 2[2AC - B^2 + kB\sqrt{2}(C - A)], \\ g_y &= -2[2AC + B^2 + kB\sqrt{2}(C + A)], \\ A^2 &+ B^2 + C^2 = 1, \end{aligned} \tag{6}$$

where k is the orbital reduction factor. This parameter, as a rule, should be in the range $0.7 \le k \le 1.0$, and it is regarded as a measure of the covalency of the metal-to-ligand bonds. The knowledge of the g values enables these equations to be solved for coefficients A, B, C, and k. The energy values (E_i) of the three Kramers' doublets, the axial (Δ) and rhombic (V) distortions are then found from the secular equations (7) in the units of the spin-orbit coupling constant (λ):

$$\begin{aligned} &(-\lambda/2 + \Delta - E)A - \frac{\lambda}{\sqrt{2}}B - 3VC = 0 \\ &-\frac{\lambda}{\sqrt{2}}A + (-2\Delta - E)B = 0 \\ &-3VA + (\lambda/2 + \Delta - E)C = 0 \end{aligned} \tag{7}$$

The paramagnetic resonance measurements give only the absolute principal g values, so that neither their signs nor the correspondence of g_1, g_2 or g_3 to g_x, g_y or g_z are known. Thus, in a common case, one must consider 48 possible combinations depending on the labeling (x, y, z) and the signs chosen for the experimental results.

A home-written program was used for all these calculations. The experimental g values were fed into the program and generated all of the possible combinations. For each set, a solution of Equations (6) for A, B, C, and k was attempted using the Newton-Raphson method. As soon as the solution was obtained, the secular Equations (7) were solved. After matrix diagonalization, the solution was checked for its correspondence to the lowest

Kramers' doublet. The best fit was selected on the basis of physically reasonable values of k, Δ and V.

The calculated parameters obtained using the above theory are listed in Table 4-4, where the assigned g values, A, B, C, k coefficients, ligand-field distortions, and the relative energies of the ground and two excited Kramers' doublets are given. It is necessary to note that only one acceptable solution exists for all of the studied Fe(III) complexes and the different types of LS centers of complex **1**. For the sake of comparison, the analogous parameters are listed in Table 4-4 for the bis(tridentate) iron(III) complex, $Fe(L)_2PF_6$, with PF_6^- as the counterion [102].

In Table 4-4 all of the iron complexes (except for LS complex **1** of the second-type) have a very large B value, which indicates that the ground-state Kramers' doublet is characterized by the state where the unpaired electron resides in the d_{xy} orbital, thus $(d_{xz},d_{yz})^4(d_{xy})^1$ is the ground state. The situation is cardinally changed for the second-type LS complex **1**. By utilizing the same axis definition, we determined g_x = 1.975, g_y = -1.98 and g_z = - 2.165, which lead to the calculated axial (tetragonal) distortion of Δ/λ = - 5.506. The negative sign indicates that d_{xy} is higher in energy than the d_{xz} and the d_{yz} orbitals. The calculated coefficients A, B, and C also indicate that the unpaired electron is localized in the $(d_{xz} \pm d_{yz})$ orbital, and thus that the $(d_{xy})^2(d_{xz},d_{yz})^3$ ground state is stabilized. It should be noted that the change between the ground states is reversible with temperature.

The possibility of the two electron configurations $(d_{xy})^2(d_{xz},d_{yz})^3$ and $(d_{xz},d_{yz})^4(d_{xy})^1$ (as shown in Scheme 4-3) for low-spin Fe(III) has previously been reported for several (porphinato)iron(III) complexes, $[TMPFe(L)_2]^+$ [99]. But the change between the ground state in these LS iron(III) complexes was caused by changing the basicity of the pyridine axial ligands: the $(d_{xy})^2(d_{xz},d_{yz})^3$ configuration was realized for the high-basicity pyridines and the $(d_{xz},d_{yz})^4(d_{xy})^1$ configuration for the low-basicity pyridines. The conversion between the $(d_{xz},d_{yz})^4(d_{xy})^1/(d_{xy})^2(d_{xz},d_{yz})^3$ configurations in a single material, namely LS iron(III) complex **1**, we observed for the first time.

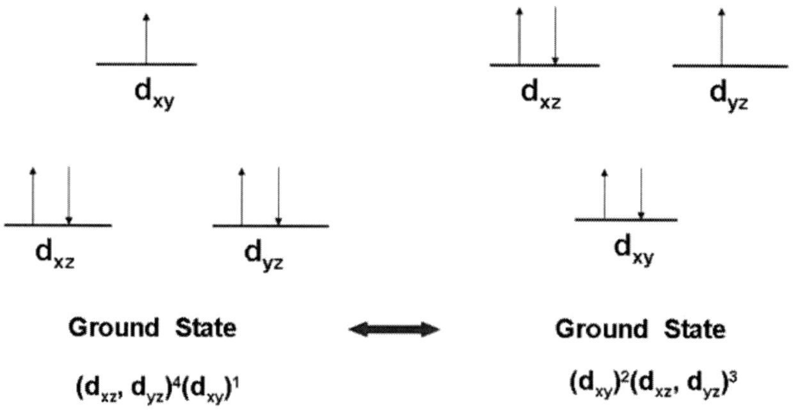

Scheme 4-3. Two possible electron configurations for the low-spin Fe(III) centers.

The dynamic process involved in changing the number of the two different types of LS centers with temperature, as illustrated in Figure 4-7a, will now be explored. The temperature dependence of increasing the number of the second-type of LS centers (n_2) from the first-type of LS centers (n_1) can be characterized by the equilibrium constant K_1 and described by Equation (4). The enthalpy (ΔH_1) and entropy (ΔS_1) of this dynamic process, ΔH_1 = 27.9 kJ/mol and ΔS_1 = 0.086 kJ/mol, were calculated from the straight line given by plotting ln K_1 vs. $1/T$ in the temperature range of 300 to 360 K (Figure 4-7b). The enthalpy value obtained is close to the activation energy E_a = 28.8 kJ/mol that was found for the reorientation of the PF_6^- ion in polycrystalline $NaPF_6$ compound [100], which was determined from the temperature dependence of the spin-lattice relaxation time T_1. The comparison between these energies is possible because the chemical bond in both $NaPF_6$ and between Fe(III) and PF_6^- in complex **1** is ionic. Thus, we postulated that the conversion between the $(d_{xz},d_{yz})^4(d_{xy})^1/(d_{xy})^2(d_{xz},d_{yz})^3$ configurations is most probably caused by the reorientation of one coordinated PF_6^- ion from the equatorial (ONN) planar position to the axial one, which was previously occupied by water.

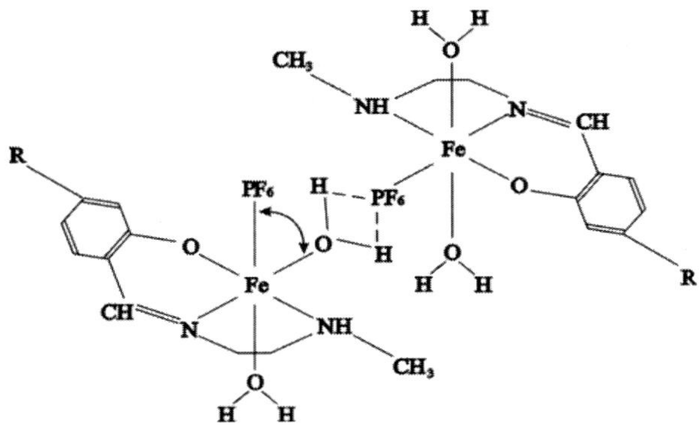

Scheme 4-4. Schematic representation of the proposed dimeric structure for iron(III) complex 1.

The EPR spectrum for complex **1** exhibited the appearance of the second-type of HS centers: a low-field broad line (600 G) with an effective g value of $g_2 \sim 5.6$ and a very broad unresolved signal in the g ~ 2 region (see Figure 4-6). This spectrum was analyzed by the rhombogram method [35, 36, 101] and indicated that the observed HS signals arise from the middle Kramers doublet when the crystal field parameters are $D > 0.3$ cm^{-1} and $E/D \sim 0.1$. It is interesting to note that the intensity of each type of LS and HS center changes synchronously: at low temperature the first-type predominates (LS center with $g_{x,y} = 2.215$, g_z = 1.935, and HS center with $D > 0.3$ cm^{-1}; $E/D \sim 0.3$), whereas at high temperature the second-type predominates (LS with $g_x = 1.975$, $g_y = 1.98$, $g_z = 2.165$, and HS with $D > 0.3$ cm^{-1}; $E/D \sim 0.1$). This synchronous behavior additionally confirmed the dimeric structure of complex **1** where the type of iron dimer is determined by the nature and location of the bridging units (H_2O or PF_6^-) between the LS and the HS centers. The proposed dimeric structure of complex **1** is schematically presented in Scheme 4-4.

4.3. Möessbauer Spectroscopy for Complex 1

In order to confirm the EPR results, complex **1** was investigated by Mössbauer spectroscopy. Mössbauer quadrupole splitting can also be used to gain information concerning the electronic ground states. The gamma-resonance spectrum obtained at 297 K (Figure 4-8) is a superposition of two broad quadrupole doublets. The more intensive doublet **1** (with an isomer shift of $\delta = 0.34$ mm/s with respect to α-Fe and with quadrupole splitting $|\Delta E_Q| = 0.82$ mm/s) has parameters associated with the low-spin ($S = 1/2$) state of the Fe(III) ion in an octahedral environment. Doublet **2** has parameters ($\delta = 0.28$ mm/s and $|\Delta E_Q| = 0.27$ mm/s) that correspond to the high-spin ($S = 5/2$) state of Fe(III) in an octahedral environment.

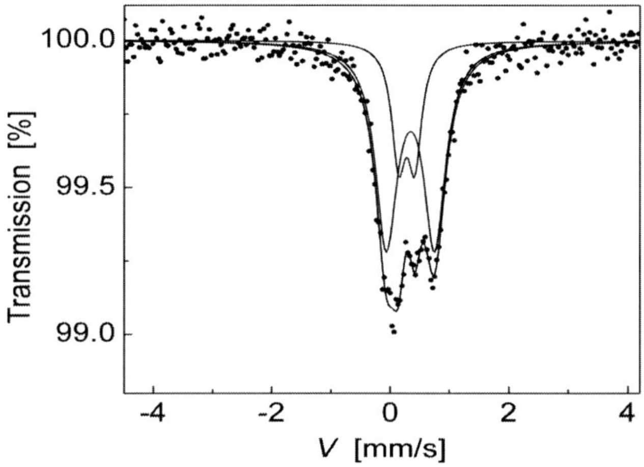

Figure 4-8. The Mössbauer spectrum for complex 1 recorded at 297 K.

At a temperature of approximately 80 K, the form of the transmission spectrum lines indicated the start of the magnetic phase transition in the sample (Figure 4-9).

The four nonequivalent positions of ^{57}Fe atoms are detected in the experimental spectrum at 5 K (Figure 4-10f). Two of the positions for the LS Fe(III) centers gave two doublets that were characterized by the following parameters: $\delta = 1.18$ mm/s, $|\Delta E_Q| = 1.2$ mm/s (Figure 4-10a) and $\delta = 0.8$ mm/s and $|\Delta E_Q| = 1.75$ mm/s (Figure 4-10b), respectively. Two of the HS Fe(III) centers showed magnetic *hyperfine structure* with the following parameters: *Zeeman sextet* with $\delta = 0.53$ mm/s, $\Delta E_Q < 0.02$ mm/s, and $H_{hf} = 519.2$ kOe (Figure 4-10c) and a typical superparamagnetic line shape for metallocomplexes with dynamics in the magnetic subsystem [103, 104] $\delta = 0.57$ mm/s, $\Delta E_Q = -0.18$ mm/s, and $H_{obs} = 456.4$ kOe (Figure 4-10e). The curve e in Figure 4-10 was obtained by subtraction of the superposition curve, d = a + b + c, from the experimental spectrum (Figure 4-10f). As the system progresses from a "slow" to a "fast" intramolecular dynamic process as the temperature increased, two of the LS and two of the HS parameters began to merge and a reduction in the number of nonequivalent positions from four at 5 K to two at 297 K was observed. The averaged quadrupole splitting for the LS centers of the Fe(III) ion was ca. 1.47 mm/s.

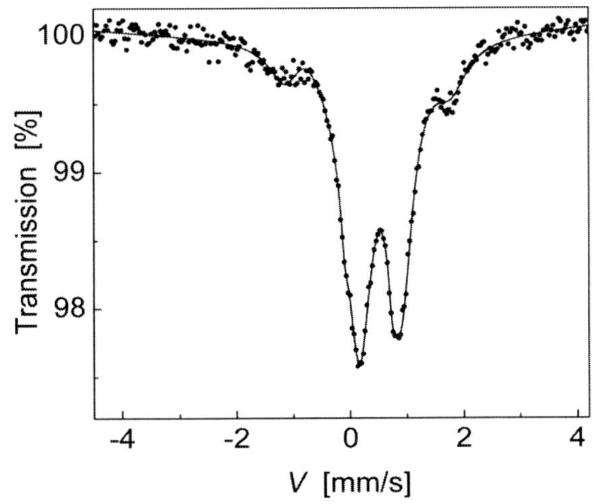

Figure 4-9. The Mössbauer spectrum for complex 1 recorded at 80 K.

The same dynamic process was registered by both EPR and Mössbauer spectroscopy but over different time scales: the time scale for the Mössbauer spectroscopy was 10^{-7} s, whereas the time scale for the EPR spectroscopy was 10^{-9} s.

Thus, the result of the thermo-stimulated intramolecular dynamics, which lead to the appearance of the time-dependent contributions in Hamiltonian of the hyperfine interactions, became visible only during the Mössbauer time frame and was observed at lower temperatures than in the EPR spectra.

The magnitude of the quadrupole splitting for the low-spin state is related to the electronic asymmetry about the ^{57}Fe nucleus. The electric field gradient QV_{ZZ} and the asymmetry parameter η are given by Equation (8) [105]:

$$\begin{aligned} QV_{ZZ}/4 &= (A^2/2+B^2/2-C^2)(1.5mm/s) \\ \eta = (V_{XX}-V_{YY})/V_{ZZ} &= -(3/2)(A^2-B^2)/(A^2/2+B^2/2-C^2) \end{aligned} \tag{8}$$

A, *B*, and *C* are the orbital coefficients defined above. The quadrupole splitting was determined by Equation (9) [106]:

$$\Delta E_Q = (QV_{ZZ}/2)(1+\eta^2/3)^{1/2} \tag{9}$$

If the *A*, *B*, and *C* coefficients are known from the EPR results, we can calculate the quadrupole splitting. From the orbital coefficients given in Table 4-4, Equation 8 and 9 give ΔE_{Q1} = - 1.40 mm/s and ΔE_{Q2} = 0.75 mm/s for the LS iron(III) centers of the first- and second-types, respectively. The calculated and observed values for the quadrupole splittings at low ($|\Delta E_Q|$ = 1.47 mm/s) and room ($|\Delta E_Q|$ = 0.82 mm/s) temperatures were in agreement.

Thus, the Mössbauer data confirm the existence of the conversion effect of the electron configurations $(d_{xz},d_{yz})^4(d_{xy})^1/(d_{xy})^2(d_{xz},d_{yz})^3$ on the basis of the quadrupole splitting values.

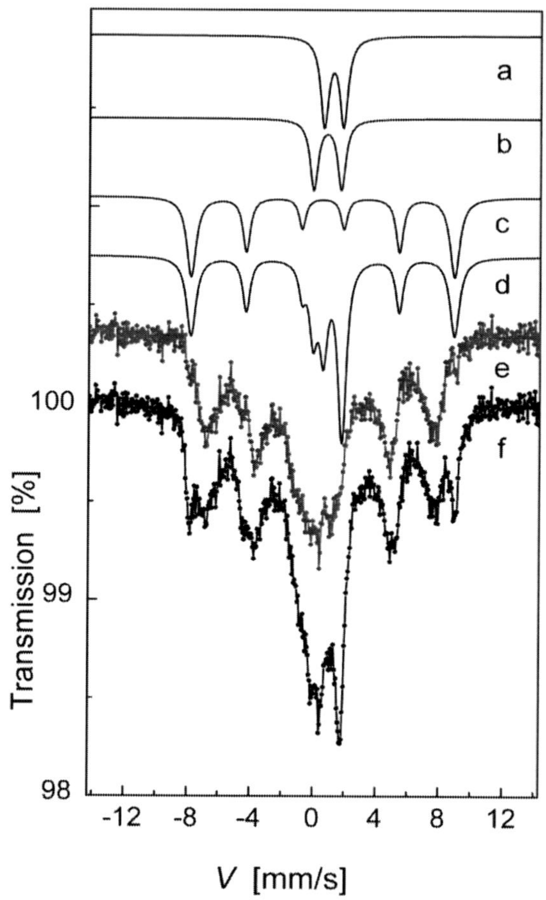

Figure 4-10. The experimental Mössbauer spectrum for complex 1 taken at 5 K (f). The solid curves (a, b, c, and d), which are located above the experimental spectrum, show the two LS and two HS iron(III) centers contributions (see text).

4.4. Summary

EPR and Mössbauer spectroscopy were successfully applied to determine the structure, magnetic, and dynamic properties of the liquid crystalline iron(III) Schiff base complexes with asymmetric ligands [107]. The six-coordinated Fe(III) complexes, formed by coordination of the three ligator (ONN) atoms of the tridentate ligand (L), the water molecules and the different (PF_6^- (**1**), NO_3^- (**2**), Cl^- (**3**)) counterions, were studied for the first time.

It was shown that all of the complexes investigated consist of low-spin (LS) and high-spin (HS) iron centers. The orienting effect of a liquid-crystalline matrix, and theoretical analysis of the observed *g* values were used in order to determine the directions of the principal (magnetic) *g* axes with respect to the molecular symmetry axes, the axial and rhombic field parameters and the relative energies of the excited and the ground states in the LS iron(III) complexes. It was determined that LS iron(III) complexes **2, 3** and LS complex **1** in the temperature range 4.2-250 K have a $(d_{xz}, d_{yz})^4(d_{xy})^1)$ ground state.

Polycrystalline monocationic Fe(III) complex $[Fe(L)X(H_2O)_2]^+X^-$ (**1**) with X = PF_6^- as the counterion, has some interesting and important features. EPR spectroscopy showed that the LS and the HS centers are coupled together in **1** and form a dimer structure by means of the water molecules and the PF_6^- counterion. The appearance of the second-type of LS and HS iron centers was risible by means of EPR spectroscopy above room temperature. This type of iron centers was best observed in the liquid crystalline (387 – 405 K) phase. The investigation of the dynamics of the appearance of the second-type of iron centers showed that they arise due to the reorientation of the PF_6^- counterion in the dimer from the equatorial (ONN) planar position to the axial one, which was previously occupied by water. Theoretical analysis of the *g* values of the second-type of LS complexes demonstrated that in this case $(d_{xy})^2(d_{xz},d_{yz})^3$ state is the ground state. Conversion between the $(d_{xz},d_{yz})^4(d_{xy})^1/(d_{xy})^2(d_{xz},d_{yz})^3$ ground states is reversible with temperature. The observed conversion took place in a single material and was detected in the iron(III) complexes for the first time. It was also shown that the type of iron dimer is determined by the type and arrangement of the bridging units (H_2O or PF_6^-) between the LS and the HS iron centers.

Thus, we have synthesized for the first time a novel compound that has a labile LS electron configuration that switches between the $(d_{xz},d_{yz})^4(d_{xy})^1/(d_{xy})^2(d_{xz},d_{yz})^3$ ground states and is driven by the LC properties. Mössbauer data confirmed the existence of such an effect on the basis of the values of the quadrupole splitting for the electron configurations $(d_{xz},d_{yz})^4(d_{xy})^1$ and $(d_{xy})^2(d_{xz},d_{yz})^3$.

5. LC Azomethine Iron(III) Complex Based on Aminopyridine

Increased interest in the studies of variable-spin transition metal complexes having switchable magnetic properties and being promising materials for molecular nanoelectronics is currently being observed [108]. These compounds have energetically close electronic terms with different spin multiplicity, which are switched by external thermal, baric, and light actions.

In iron(III) complexes with an octahedral coordination core Fe(III) ions are either in high-spin ($^6A_1$, $S = 5/2$) or low-spin ($^2T_2$, $S = 1/2$) states depending on the ratio between the splitting values of *d* electron energy levels in the cubic crystal field and the spin coupling energy. The Fe(III) complex with a Schiff base (H_{pap} ligand) is known to exhibit a spin-crossover ($S = 1/2 \leftrightarrow 5/2$) property and under irradiation (with a wavelength of 300 nm) at a temperature of 5 K it demonstrates the LIESST effect [109]: a transition from the low-spin (LS) state to a metastable high-spin (HS) state.

It seemed interesting to find out how the liquid crystal properties of the compound affected its magnetic variable-spin behavior and the occurrence of the LIESST effect. To this end, we synthesized a new liquid crystal iron(III) complex with a Schiff base (4,4'-dodecyloxybenzoyloxybenzoyl-4-oxysalicylidene-2-aminopyridine with a PF_6^- counterion) having the same coordination sphere and the ligand structure as in the $Fe(pap)_2ClO_4{\cdot}H_2O$ complex with the H_{pap} ligand, but involving long linear substituents providing the liquid crystal properties of the compound.

5.1. Synthesis, Characterization of Iron Complex and Identification of the LC Properties

Synthesis of the target product was carried out directly through the formation of Shiff base without isolation of the ligand from the solution and the subsequent complex formation with the salt of iron(III). The substitution of the counterion was carried out through the exchange reaction by adding a double excess of the salt.

The structure of the complexes was established by infrared (IR) vibrational spectroscopy. The formation of the azomethine takes place in the course of the reaction between the aldehyde and 2-aminopyridine directly in solution as confirmed by the presence of a strong band at 1674 cm^{-1} characteristic of the HC=N bond [110,111], which lies close to the absorption band of the carboxyl group C=O (1733 cm^{-1}) of the ester fragment of the molecules [112,113].

Analysis of infrared spectra in the far range reveals that in the nearest environment the iron ion is coordinated by two hydroxyl groups (a fragment of salicylic aldehyde). That is confirmed by the presence a number of vibrations bands of the Fe-O bond (408 – 397 cm^{-1}) formed due to the substitution of a proton [114-116]. The absence of the signals of proton from hydroxyl group of the aldehide in NMR spectra (9.9 ppm) also indicates on formation Fe-O bond [117]. The coordination plane around iron takes two nitrogen atoms of the azomethine (Fe-N bond, 506-499 cm^{-1}) [118]. The formation of the octahedral packing of iron is finished by the Fe-N_{py} bonds which are in the axial positions above and under the plane (356-330 cm^{-1}) [119]. A similar coordination of pyridine through the heteroatom with a metal ion is also found to (oxo)(hydroxo) molibdenumtetraphenilporphine [120].

The absorption band of the counterion is specific in the complex ~ 846 cm^{-1} [111,121].

The stability and the presence of iron in the structure of complex were confirmed by the MALDI-ToF MS data [122,123]. Table 5-1 gives the data of the mass-spectra of complex which indicate the existence of a series of stable ions characterizing the iron(III) ion in its immediate environment.

Table 5-1. Value of m/z of the complex

Ion	m/z	
	Experimental	Theoretical
$C_{76}H_{82}N_4O_{12}Fe\cdot PF_6$		
$[L]^+$	-	622.76
$[L\cdot K]^+$	668.47	663.77
$[2L\cdot Fe\cdot 2Li]^+$	1315.25	1315.25

Based on the elemental analysis and mass spectrometric data the following possible model of the iron(III) complex can be proposed(Figure 5-1).

Liquid crystal phase transitions of the compound were examined by differential scanning calorimetry (DSC) and polarized thermal microscopy which showed that the iron(III) complex exhibited mesomorphic properties. The values of temperatures and enthalpies of the phase transitions are listed in Table 5-2.

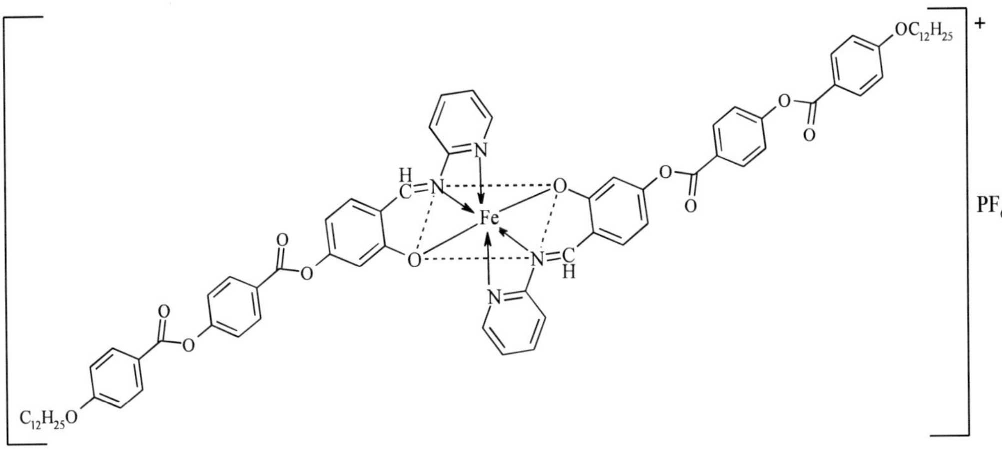

Figure 5-1. Schematic model of the complex.

Table 5-2. The phase transition temperatures according to the DSC in the heating cycle

Compound	T_g [°C]	ΔCp [J/g·K]	T_{ph1} [°C]	ΔH_1 [J/g]	T_{ph2} [°C]	ΔH_2 [J/g]	T_m [°C]	ΔH_m [J/g]
$C_{76}H_{82}N_4O_{12}$ Fe·PF_6	18.7	0.39	83.1	*	93.2	48.93*	136.4	6.54

Thus, in the heating cycle at 66.5°C a transition from the solid state to non-geometric texture occurs followed by the formation of myelin (smectic) pathways at 97 °C. The clearing point is achieved at 137 °C

$$\mathrm{Cr} \rightarrow 66{,}5^{\circ}\mathrm{C} \xrightarrow{\mathrm{Mes}} 97\,^{\circ}\mathrm{C} \xrightarrow{\mathrm{Sm}} 137^{\circ}\mathrm{C} \rightarrow \mathrm{Iso},$$

where *Cr* is the crystalline state and Iso is the isotropic state.

Figure 5-2 depicts the sample texture in the heating cycle (nicols are crossed) in the transition of the non-geometric texture to the smectic (Sm) phase with an increase in myelin pathways.

5.2. EPR Characterization of LC Azomethine Iron(III) Complex Based on the Aminopyridine

The EPR spectra of the compound (X-range, $h\nu$ = 0.3 cm^{-1}) were measured in the temperature range 4.2 – 300 K and the most characteristic spectra are depicted in Figure 5-3.

At low temperatures the spectrum contains two main signal: a strong intensity line with g_1 = 4.3, increasing with a decrease in the temperature, on whose left wing an additional line with g_2 = 9.6 appears and increases as the temperature decreases and a weak intensity line with the effective g factor g_3 = 2 and the width $\Delta H_{p\text{-}p}$ = 200 G. Above 100 K, a new broad

($\Delta H_{p\text{-}p}$ = 2450 G) structural line with the *g* factor close to 2 starts to appear in the spectrum, which grows in intensity as the temperature increases.

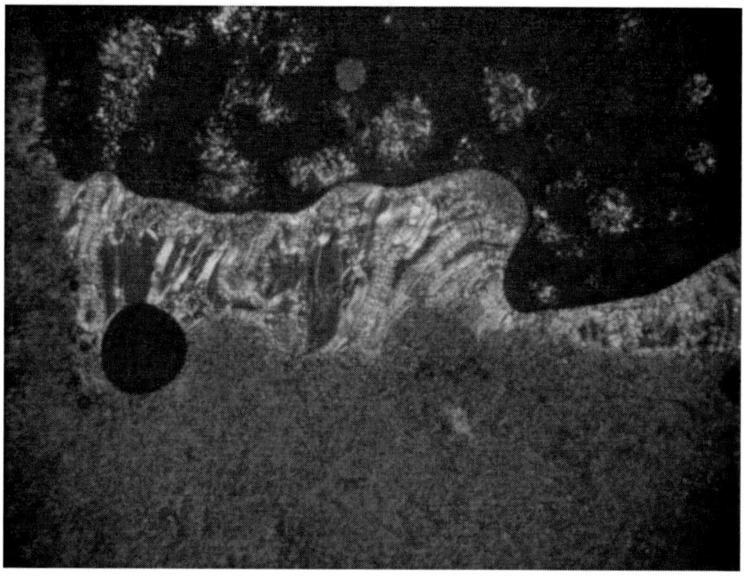

Figure 5-2. Transition of non-geometric texture to the smectic phase with the growth of myeline paths, T = 98°C, nicols are crossed.

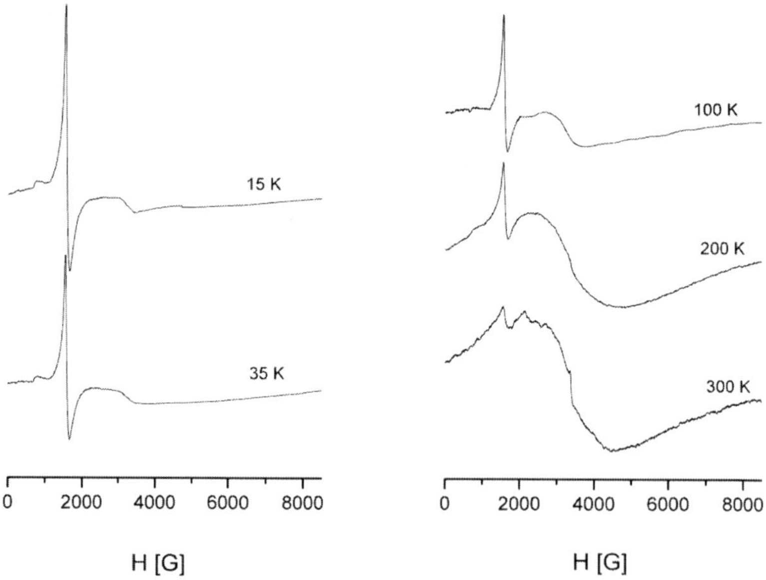

Figure 5-3. EPR spectra (X-range) of the compound at different temperatures.

The low-temperature spectrum (T < 80 K) belongs to HS Fe(III) ions with the $^6A_{1g}$ ($3d^5$) ground state and is described by the spin-Hamiltonian (1).

Theoretical calculations of the low-temperature spectrum show [35, 36] that signals with $g_1 = 4.3$ and $g_2 = 9.6$ characterize the HS iron centers with the fine structure parameter $D >> h\nu = 0.3$ cm^{-1} and a strong rhombic distortion ($E/D = 1/3$) (**I**-type centers). The signal with the effective g factor $g_3 = 2$ is due to the HS iron(III) centers with a high symmetry ($D << h\nu = 0.3$ cm^{-1}) crystal field on the ion (**II**-type centers). The fraction of **II**-type complexes is small and according to the ratio of the integrated intensities of EPR lines it is 11 %.

The study of the temperature dependence of the integrated intensity of the EPR spectrum (I) of the compound shows that it is complex and consists of three steps: passes through the maximum at 15 K, then monotonically decreases to a minimum value at 80 K, and then gradually starts to grow (Figure5-4a). It is known that the product of the integrated intensity of the EPR spectrum on the temperature (I•T) for mononuclear paramagnetics must not depend on the temperature. For the compound being studied the stepwise behavior of this value is observed (Figure 5-4b).

At present, for iron complexes three possible reasons for the stepwise dependence of the product (I•T) (which is proportional to χ•T, where χ is the magnetic susceptibility) on the temperature have been known [124]: the formation of binuclear compounds; the occurrence of specific antiferromagnetic interactions between the LS and HS centers, resulting in the stepwise behavior of the spin transition [125,126]; the presence of two different spin transitions in the system occurring at different temperatures [108]. Since no spin transition is observed in the system under study (only HS centers are present), then the possible reason for the stepwise behavior of the (I•T) value on T is the formation of binuclear compounds. Note that a similar stepwise temperature dependence for the product of the magnetic susceptibility on the temperature (χ•T) was detected for binuclear μ-oxo-complexes of iron(III) (Figure 11, ref. [127]), in which the inflection point was observed in the curve at 75 K, approximately as in our case. It is also known that binuclear μ-oxo-complexes of iron(III) are characterized by a strong intramolecular antiferromagnetic exchange interaction ($H = JS_1S_2$) through the Fe- O - Fe oxygen bridge, and the exchange value varies $J \approx 160 \div 265$ cm^{-1} depending on ligands [128].

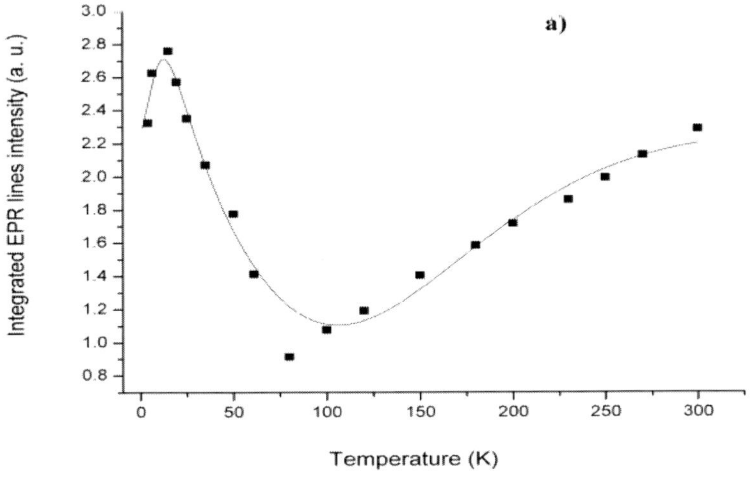

Figure 5-4. (Continued).

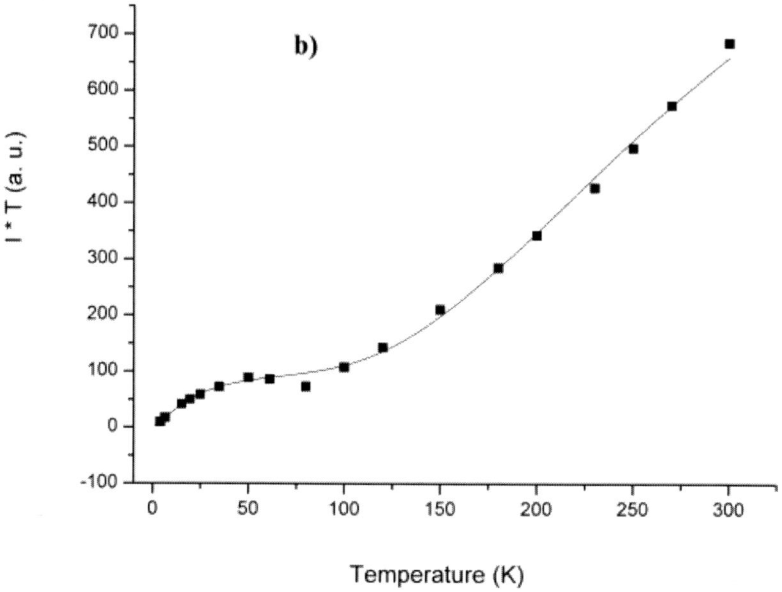

Figure 5-4. Temperature dependence of the integrated EPR spectrum intensity of the iron complex (a). Temperature dependence of the product (I•T) (b). Solid line is calculated by equations (11) and (12) (see text), dots show experimental values.

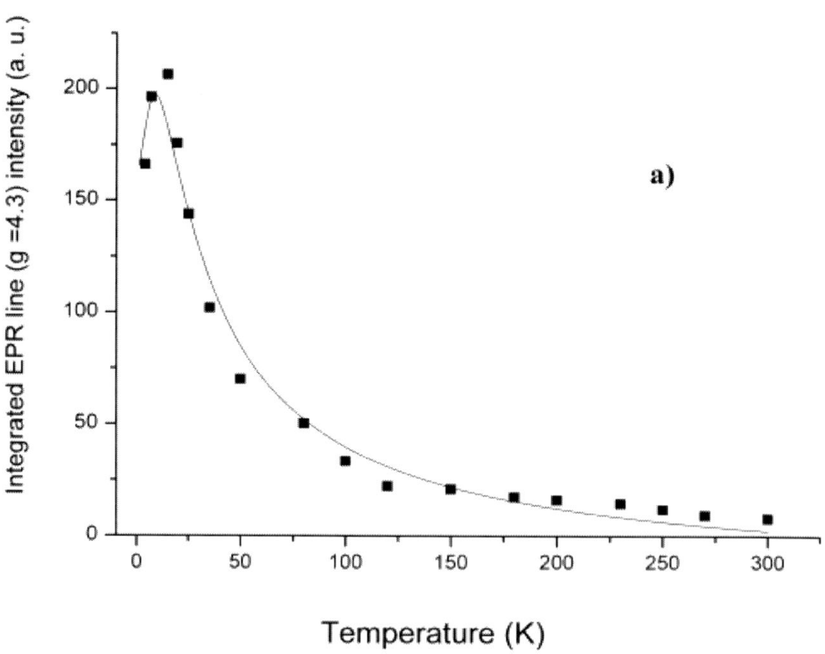

Figure 5-5. (Continued).

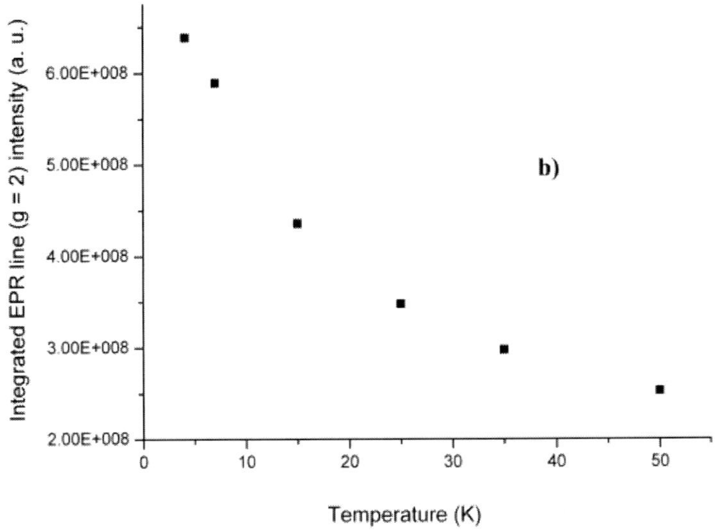

Figure 5-5. Temperature dependence of the integrated EPR line intensity for **I**-type centers (a). Solid line calculated by equation (11). Temperature dependence of the integrated EPR line intensity for **II** – type centers (b).

In order to reveal the exchange interactions in our compound we analyzed separately the temperature dependence of the integrated intensity of EPR lines for each type of magnetic centers. This dependence for the **I**-type of HS centers is shown in Figure 5-5a. It demonstrates a maximum at T_N = 15 K; above T_N the number of centers monotonically decreases. The presence of the maximum in the curve can be due to two reasons: exchange interactions between the magnetic centers and the magnetic anisotropy of the Fe(III) ion (parameter D).

The contribution from the magnetic anisotropy of the Fe(III) ion to the magnetic susceptibility (proportional to the integrated intensity of EPR lines) is defined [129] by the expression:

$$\chi = (\chi_{\parallel} + 2\chi_{\perp})/3,$$
$$\chi_{\parallel} = \frac{C(1+9e^{-2b}+25e^{-6b})}{4(1+e^{-2b}+e^{-6b})}, \qquad (10)$$
$$\chi_{\perp} = \frac{C[9+8(1-e^{-2b})/b+9(e^{-2b}-e^{-6b})/2b]}{4(1+e^{-2b}+e^{-6b})},$$

where $b = D/kT$ and $C = Ng^2\mu^2_B/kT$. A calculation by formulas (10) shows that the best agreement with the experimental data is obtained at D = 2 cm^{-1}. However, for this D value, according to theoretical calculations [130], in the X-range ($h\nu$ = 0.3 cm^{-1}) quite another EPR spectrum must be observed. Thus, the appearance of the maximum in the curve is due to the occurrence of the exchange interactions. This is also supported by the results of Mössbauer spectroscopy presented in subsection 5-3 (Figure 5-11). The calculation of the temperature dependence of the integrated intensity of EPR lines for the **I**-type centers was performed within a model of linear chains with the isotropic exchange interaction ($H = 2J\sum_{nn} S_i S_j$) because the chain structure facilitates the packing of molecules into smectic layers and our compound demonstrates the presence of the smectic phase at high temperatures. The analytical function describing the temperature dependence of the magnetic susceptibility is defined by the following expression [131,132]:

$$\chi = \frac{NS(S+1)g^2\mu_B^2}{3kT}\frac{1+u(K)}{1-u(K)};$$

$$\text{where: } u(K) = \coth K - 1/K; \tag{11}$$

$$K = -\frac{2JS(S+1)}{kT}, S = 5/2,$$

and $J > 0$ corresponds to antiferromagnetic exchange interactions. The processing of the temperature dependence (Figure 5-5a) of the integrated intensity for the **I**-type centers by formula (11) gives the exchange integral $2J = 1.5\ \text{cm}^{-1}$.

The study of the temperature dependence of the intensity of EPR lines of HS **II**-type centers shows (Figure 5-5b) that it does not have a maximum, and is described by the Curie law, i.e. the **II**-type centers are isolated and do not participate in the exchange interactions. The fraction of these centers is small (~ 11%).

Now let us consider the behavior of the high-temperature EPR spectrum ($T > 80$ K), namely, the behavior of a broad structural line with the g factor of 2, which increases in intensity as the temperature increases. As noted above, the identity of the magnetic behavior ($\chi \cdot T$) depending on T for binuclear μ-oxo-complexes of iron(III) [127] and ($I \cdot T$) for our compound allows us to suppose that above 80 K the iron complex being studied has a dimeric structure due to a strong antiferromagnetic exchange interaction through oxygen bridges. This assumption is also confirmed by the analogy in the behavior of the EPR spectra of the binuclear μ-oxo-complex of iron(III) [133] and our compound. In the EPR spectrum of the binuclear iron(III) complex, as in our case, on the base line with $g = 4.3$ above 80 K a new spectrum appears with a center $g \sim 2$, which grows in intensity as the temperature increases [133]. In ref. [127, 133] there are X-ray crystallographic data for single crystalline samples, which visibly demonstrate the formation of dimers through the oxygen bridge.

In this context, we calculated the temperature dependence of the integrated intensity of a broad structural line for the **III**-type centers (Figure 5-6). In the case of a strong antiferromagnetic ($J > 0$) exchange interaction $J(S_1S_2)$ coupling two ions with spin $S_1 = S_2 = 5/2$, the isotropic exchange results in the formation of ($2S_1 + 1$) multiplets, where the singlet ($S = 0$) nonmagnetic state is the ground state and the excited states with $S = 1, 2, 3, 4, 5$ are located higher by J, $3J$, $6J$, $10J$, and $15J$. At high temperatures the excited multiplets are populated, which determines the appearance of new lines in the EPR spectrum corresponding to $S = 1, 2, 3, 4, 5$ states. The temperature variation of the EPR signal intensity for each states populated multiplets be described by the relationship [134,135]:

$$I \sim \frac{\exp\left(\frac{-S(S+1)J}{2kT}\right)}{T\sum_{S=0}^{5}(2S+1)\exp\left(\frac{-S(S+1)J}{2kT}\right)}. \tag{12}$$

The calculation of the temperature dependence of the integrated EPR signal intensity allows us to find out to which thermally populated multiplet the observed spectrum is due and provides the possibility to determine the isotropic exchange value. As seen (Figure 5-6), the temperature variation of the integrated intensity of the new broad structural line is well described by the magnetic excited state with $S = 2$ in accordance with expression (12) with the isotropic exchange value $J = 199$ cm^{-1}. Note that the examples of the assignment of the observed EPR spectrum only to the excited state $S = 2$ for iron(III) dimers with oxygen bridges have previously been described in the literature [135,136].

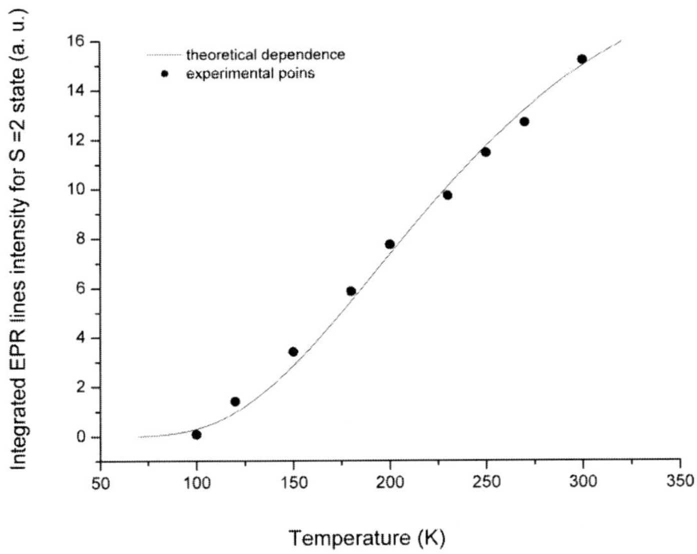

Figure 5-6. Temperature dependence of the integrated EPR line intensity for the III type centers.

Now let us attempt to present in general the structural organization of the system being studied. The results show that at a low temperature ($T < 80$ K) monomeric iron complexes are organized in 1D linear chains, and above 80 K the chains consist of dimeric molecules whose metal centers are linked by a strong antiferromagnetic exchange interaction through oxygen bridges. Within this concept let us calculate the temperature dependence of the integrated EPR line intensity of the whole spectrum (I) (Figure 5-4a), describing this behavior by a sum of equations (11) and (12) and using the previously found values $J_{01} = 199$ cm^{-1} for **III** and $J_{02} = 1.5$ cm^{-1} **I** type centers as zero approximations. Thus obtained theoretical dependences I and products ($I \cdot T$) together with experimental values are given in Figure 5-4a and 5-4b, respectively. They were calculated with the exchange constants $J_1 = 162.1 \pm 0.5$ (cm^{-1}) for the dimer and $J_2 = 2.1 \pm 0.5$ (cm^{-1}) for the chain. Good agreement between theory and experiment is observed in the whole temperature range (4.2 – 300 K). A small difference is found only at T = 80 K, which is likely to be explained by that at this temperature the structural phase transition (a transition from a 1D chain organization of HS Fe(III) centers to dimeric molecules) occurs in the system.

The question arises: how dimeric molecules form in our system? The answer to it we obtained by means of quantum chemical (DFT) calculations.

Quantum Chemical Calculations

Primarily, we calculated a monomeric complex that can have different structures of the coordination core because the ligand having sufficient flexibility can coordinate differently. Indeed, for the LS state we managed to localize the equilibrium structures of all four isomers: *trans/trans*, *trans/cis*, *cis/trans*, *cis/cis*, where *cis* or *trans* corresponds to the orientation of two phenolate oxygen atoms and two pyridine nitrogen atoms (Figure 5-7). The ligand fragment is bent in the first three coordination models (O and N_{py} atoms are in the *cis* position relative to each other). On the contrary, the ligand fragment is planar in the *cis/cis* isomer (O and N_{py} atoms are in the *trans*-position relative to each other). Note that for the HS state we found only one minimum on the potential energy surface corresponding to the *cis/cis* isomer. It is important, however, to emphasize that only the *trans/trans* model shown in Figure 5-7 provides the antiparallel orientation of long ligand substituents, and hence, it is the only coordination model consistent with the molecular packing in the smectic phase and X-ray diffraction data presented below. Thus, although the *cis/cis* isomer is the most stable in the gas phase, the crystalline and liquid crystal phases appear to be formed by the *trans/trans* monomer (T < 80 K) and its dimer (T > 80 K).

At the next stage we pass to the consideration of the dimer whose formation is detected by the EPR spectra at T > 80 K. We managed to localize its equilibrium structure in the HS state. In the optimized structure (Figure 5-8), the coordination polyhedra of two metal centers are strongly distorted, but structurally equivalent. This coincides with the Mössbauer spectroscopy results (Figure 5-9). On each metal center the pyridine ring, whose nitrogen atom must complete the coordination sphere, is almost discoordinated; another pyridine ring is discoordinated due to the formation of a new Fe–O bond with the oxygen atom of the second binuclear iron center. According to our calculations, metal centers ($S_1 = S_2 = 5/2$) are coupled by the antiferromagnetic exchange interaction with $J = 19$ cm^{-1}. A smaller exchange parameter in comparison with the value obtained from the EPR data can be explained by the packing effects. The crystal structures are always more compact and compressed, and consequently, are characterized by a stronger exchange interaction.

5.3. Möessbauer Studies of the Iron Complex

The spectrum of the compound being studied at room temperature (Figure 5-9) contains a quadrupole doublet and a singlet line. The isomer shifts are 0.37 mm/s for the doublet and 0.12 mm/s for the singlet with respect to α - Fe, and the quadrupole splitting for the doublet is ΔE_Q = 0.78 mm/s. The area under the curve of the quadrupole doublet is ~89% of the total area of the Mössbauer spectrum.

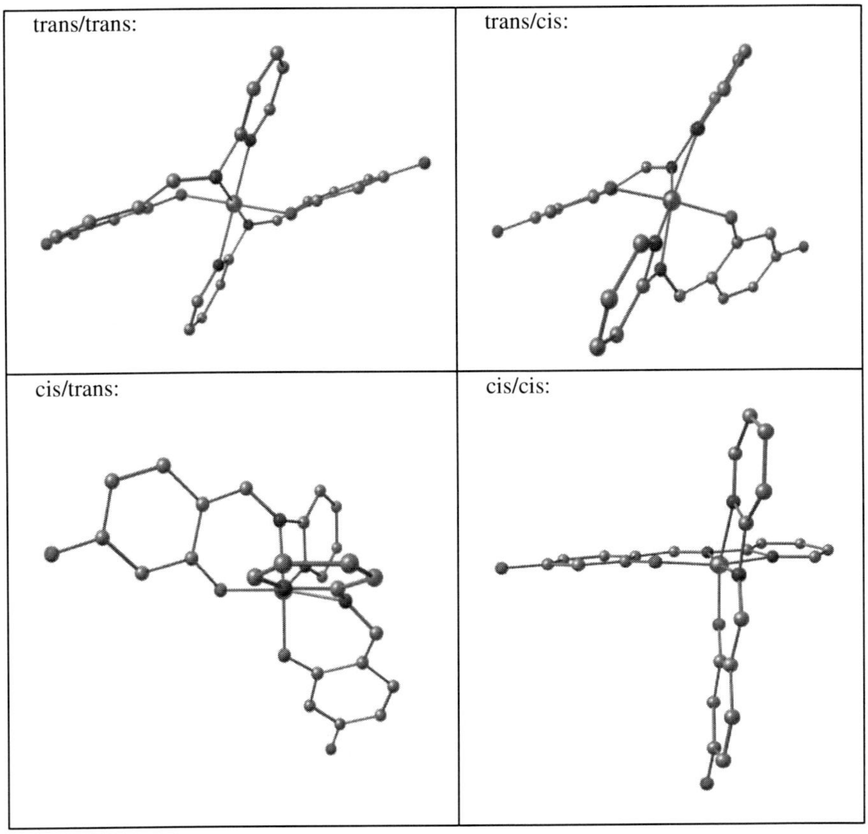

Figure 5-7. Coordination models of the monomeric iron complex (long ligand substituents are omitted).

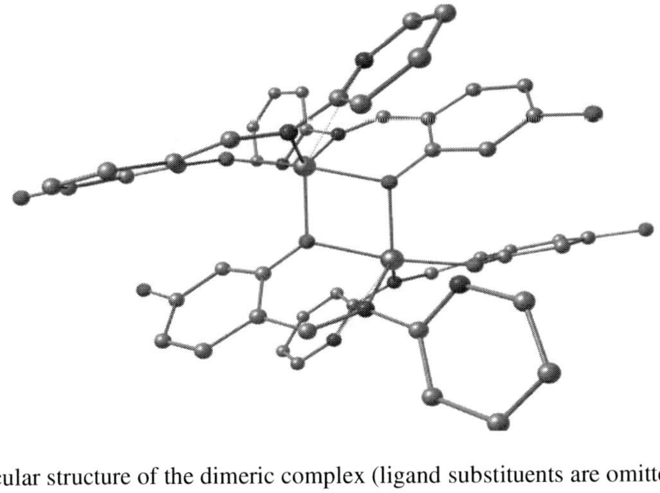

Figure 5-8. Molecular structure of the dimeric complex (ligand substituents are omitted).

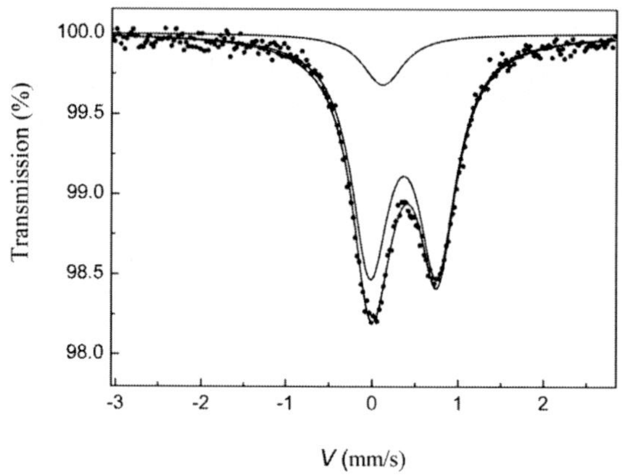

Figure 5-9. Mössbauer spectrum of the compound at 300 K.

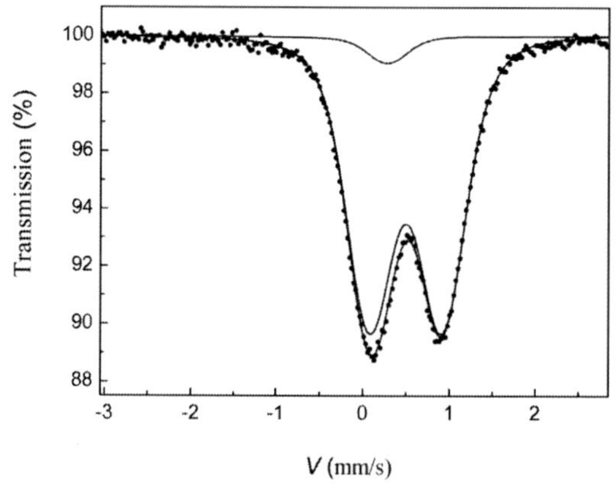

Figure 5-10. Mössbauer spectrum of the sample at 80 K.

A large value of the quadrupole splitting indicates a strong distortion of the coordination polyhedra of Fe(III) ions. At 80 K (Figure 5-10) the area of the singlet component (δ_{Fe} = 0.27 mm/s, 80 K) is ~4 %. A decrease in the singlet area is also observed when the temperature decreases further, which seems to be due only to the relative temperature course of Mössbauer-Lamb factors of the phases studied. The line shape of the doublet component at 80 K (δ_{Fe} = 0.49 mm/s, ΔE_Q = 0.85 mm/s) is not well described by the Lorentz line shape, but it is described by a mixed Lorentz-Gauss line shape with a coefficient of 0.8. This fact indicates the proximity of characteristic dynamic times in the spin subsystem to the times of the Mössbauer time window and the proximity to the magnetic phase transition point. At 5 K the magnetic hyperfine structure (H_{obs} = 436 kOe) appears in the spectrum, which gives evidence of the magnetic ordering in the system. In order to determine the spectral shapes and line positions of the hyperfine structure we applied expression (18) from ref. [137]. The magnetically ordered part of the spectrum was well described by three relaxation sextets (Figure 5-11, solid lines) with hyperfine parameters δ_{Fe} = 0.5 mm/s, ΔE_Q < 0.025 mm/s, H_{obs} = 436 kOe. Their behavior was set by expression (18) from ref. [137] and they differed only in the average relaxation times and partial areas. The found parameters of all Mössbauer spectra indicate the presence of only HS Fe(III) centers in the system.

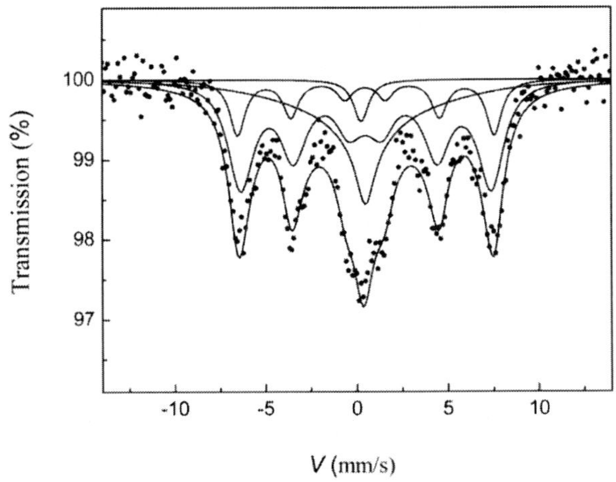

Figure 5-11. Mössbauer spectrum at 5 K.

Based on the above results it is possible to propose the following models of the molecular packing in the iron(III) complex. The high-temperature phase ($T > 80$ K) is formed by chains consisting of dimeric molecules (Figure 5-12a). In the low-temperature phase ($T < 80$ K) the chains consist of monomeric iron complexes linked with each other by the π-π stacking of pyridine rings (Figure 5-12b).

The models proposed are consistent with the structure of the molecular packing of the iron complex in the rhombic lattice, which was obtained in ref. [138] by X-ray diffraction on the polycrystalline compound (Figure 5-13). The authors of [138], however, note that the X-ray diffraction spectrum of the compound at room temperature contains maxima with very strong (especially at large scattering angles) radial smearing, which gives evidence of significant distortions in the spatial molecular packing of the iron complex. Apparently, it is the reason for that the diffraction pattern indication did not allow the authors of [138] to clearly detect the dimeric structure of iron complexes along the chain.

5.4. Summary

The results of the study of new liquid crystal iron(III) complex with a tridentate ligand of the Schiff base is reported in this section [139]. It is found out that the introduction of long linear substituents in the ligand, which provide the liquid crystal properties of the complex, significantly changes the spin-crossover behavior of the central magnetic Fe(III) ion and transforms all magnetic centers to the HS state. The employment of EPR spectroscopy allowed us to establish the magnetic structure of the compound. Unusual stepwise behavior of the product of the integrated EPR line intensity (I) and the temperature with an inflection point at ~ 80 K is found. It is shown that below 80 K iron(III) complexes are organized in 1D antiferromagnetic chains due to the π-π stacking of pyridine rings between the nearest neighbors with an exchange value of 2.1 cm^{-1}. Above 80 K the chains consist of dimeric molecules whose metal centers are bonded by a strong antiferromagnetic exchange interaction through oxygen bridges and the exchange value in the dimer is 162.1 cm^{-1}. Mössbauer spectra confirm the presence of only HS iron(III) centers in the system and the transition of the system to the antiferromagnetic ordered state at 5 K.

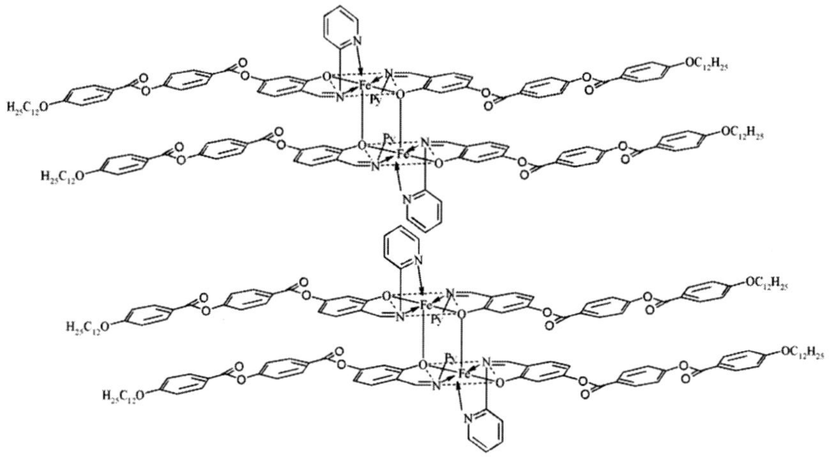

a)

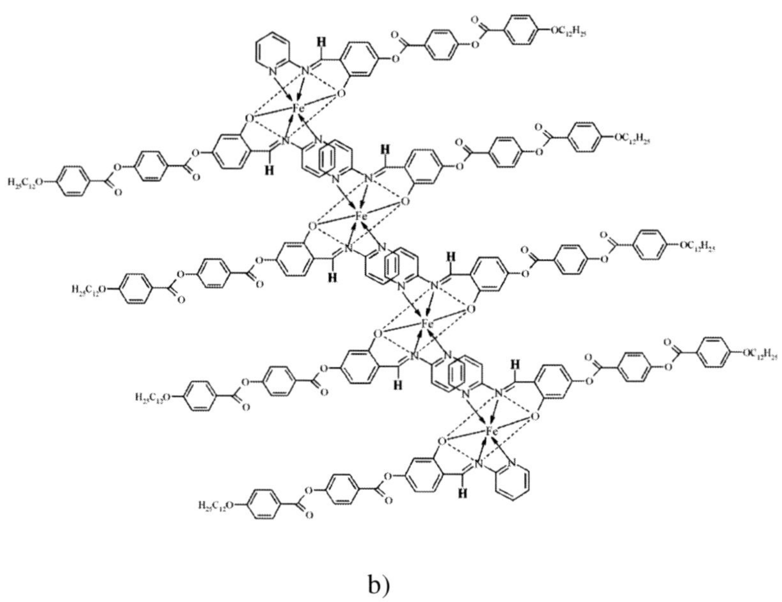

b)

Figure 5-12. Structure model of the compound at T >80 K (a) and at T <80 K (b).

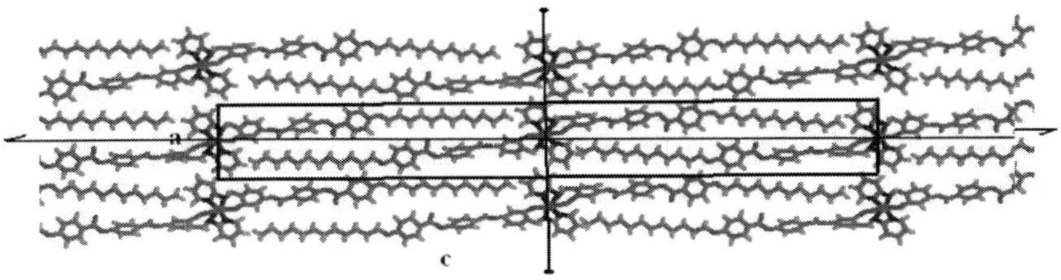

Figure 5-13. Molecular packing of the iron complex in the rhombic lattice with the space group $P2_122$ (ac cross-section).

Conclusion

The structural and magnetic properties of four different types of iron-containing liquid crystals (LC), metallomesogens, were briefly considered. EPR and Mössbauer spectroscopy were used for observing the magnetic behavior and characterization of iron-containing liquid crystals.

The first section is devoted to iron(III) complexes created on the basis of the LC poly(propylene imine) dendrimers from the first to fifth generation, which exhibit a columnar packing of the molecules in the mesophase. It has been shown that iron ions existing in a high-spin state are coordinated in dendrimer ligands by two kinds of iron-complexing sites with an octahedral and a tetrahedral symmetry. Octahedral (high-symmetry) centers are located at the border of the dendrimer, whereas the tetrahedral (rhombic distorted) centers are distributed throughout all branching of the dendrimer. The octahedral "bordering" centers are exchange-coupled and are likely to form chains.

The second section deals with γ-Fe_2O_3 nanoparticles (NPs) fabricated inside LC poly(propylene imine) dendrimeric matrix, which self-organizes into a hexagonal columnar phase. It has been found that a temperature-driven transition from superparamagnetic to ferrimagnetic state was detected by EPR, wherein the blocking temperature was about 60 K. The magnetic moment and effective magnetic anisotropy of NPs have been determined from the analyses of the EPR data. The effective magnetic anisotropy constant is found to be enhanced compared to the bulk γ-Fe_2O_3 and this enhanced value is associated with the influence of the surface and shape effects. The angular dependence of the EPR signal position for field-freezing sample from liquid-crystalline phase showed that NPs possess the uniaxial anisotropy, in contrast to bulk γ-Fe_2O_3.

The third section is devoted to LC iron(III) Schiff base complexes with asymmetric ligands. EPR showed that complexes consist of two kinds of iron centers: S = 1/2 low-spin (LS) and S = 5/2 high-spin (HS). Interesting features were found for the monocationic Fe(III) complex, $[Fe(L)X(H_2O)_2]^+X^-$, with X = PF_6^- counterion. LS and HS centers in this compound are coupled together by antiferromagnetic exchange interactions and form a dimer structure by means of the water molecules and the PF_6^- counterion. It has been shown that this novel material has a labile LS electron configuration that switches between the $(d_{xz},d_{yz})^4(d_{xy})^1/(d_{xy})^2(d_{xz},d_{yz})^3$ ground states and is driven by the LC properties. Such interesting feature was detected for the first time in iron mesogens.

The fourth section deals with LC azomethine iron(III)complex based on the aminopyridine. It has been found that the introduction of liquid crystalline substituents in the ligand, significantly changes the spin-crossover behavior of the central magnetic Fe(III) ion and transforms all magnetic centers to the HS state. Unusual stepwise behavior of the product of the integrated EPR line intensity and the temperature with an inflection point at ~ 80 K is revealed. It is shown that below 80 K iron(III) complexes are organized in 1D antiferromagnetic chains due to the π-π stacking of pyridine rings between the nearest neighbors with an exchange value of 2.1 cm^{-1}. Above 80 K the chains consist of dimeric molecules whose metal centers are bonded by a strong antiferromagnetic exchange interaction through oxygen bridges and the exchange value in the dimer is 162.1 cm^{-1}.

Acknowledgment

We gratefully acknowledge the financial support for this work by RAS Presidium program No. 24П.

References

[1] Serrano, J-L. Metallomesogens: Synthesis, Properties, and Applications; *Wiley-VCH Verlag GmbH: Weinheim,* 1996.
[2] Griesar, K.; Haase, W. Magnetic properties of transition-metal-containing liquid crystals, in Magnetic Properties of Organic Materials; editor P.M. Lahti, *Marcel Dekker:* New York, 1999.
[3] Binnemans, K.; Groller-Walrand, C. *Chem Rev.* 2002, 102(6), 2303-2346.
[4] Lee, C. H.; Kwon, Y-W.; Choi, D. H.; Geerts, Y. H.; Koh, E.; Jin, J-I. *Adv Mater.* 2010, 22(39), 4405-4409.
[5] Belarbi, Z.; Sirlin, C.; Simon, J., Andre, J. J. *J. Phys. Chem.* 1989, 93(24), 8105-8110.
[6] Barbera, J.; Iglesias, R..; Serrano, J. L.; Sierra, T.; de la Fuente, M. R.; Palacios, B.; Perez-Jubindo, M. A.; Vazquez, J. T. *J. Am. Chem. Soc.* 1998, 120(12), 2908-2918.
[7] Serrano, J. L.; Sierra, T. *Chem. Eur. J.* 2000, 6(5), 759-766.
[8] Galyametdinov, Y.; Ksenofontov, V.; Prosvirin, A.; Ovchinnikov, I.; Ivanova, G.; Gutlich, P.; Haase, W. *Angew. Chem. Int. Ed.* 2001, 40(22), 4269-4271.
[9] Fujigaya, T.; Jiang, D-L.; Aida, T. *J. Am. Chem. Soc.* 2003, 125(48), 14690-14691.
[10] Hayami, S.; Moriyama, R.; Shuto, A.; Maeda, Y.; Ohta, K.; Inoue, K. *Inorg. Chem.* 2007, 46(17), 7692-7699.
[11] Dierking, I. Textures of Liquid Crystals; *Wiley-VCH Verlag GmbH: Weinheim,* 2003.
[12] Donnio, B.; Garcia-Vazquez, P.; Gallani, J. L.; Guillon, D.; Terazzi, E. *Adv. Mater.* 2007, 19(21), 3534-3539.
[13] Alonso, P. J. Electron Paramagnetic Resonance of Paramagnetic Metallomesogens, in Metallomesogens: Synthesis, Properties, and Applications; editor Serrano, J.L.; *Wiley-VCH Verlag GmbH:* Weinheim, 1996.
[14] Galyametdinov, Y. G.; Knyazev, A. A.; Dzhabarov, V. I.; Cardinaels, T.; Driesen, K.; Görller-Walrand, C.; Binnemans, K. *Adv. Mater.* 2008, 20(2), 252-257.

[15] Piguet, C.; Bunzli, J-C. G.; Donnio, B.; Guillon, D. *Chem. Commun.* 2006, 36, 3755-3768.
[16] Domracheva, N. E.; Ovchinnikov, I. V.; Turanov, A. N.; Konstantinov, V. N. *J. Magn. Mater.* 2004, 269(3), 385-392.
[17] Ovchinnikov, I. V.; Ivanova, T. A.; Petrashen, V. E.; Galyametdinov, Yu. G.; Ivanova, G. I. *Appl. Magn. Reson.* 2005, 29(2), 325-334.
[18] Balzani, V.; Campagna, S.; Denti, G.; Juris, A.; Serroni, S.; Venturi, M. *Acc. Chem. Res.* 1998, 31(1), 26-34.
[19] Diederich, F.; Gross, M.; Knobler, C. *Angew Chem Int Ed Engl.* 1994, 33(17), 1739-1742.
[20] Tomoyose, Y.; Jiang, D. L.; Jin, R. H.; Aida, T. *Macromolecules* 1996, 29(15), 5236-5238.
[21] Moulines, F.; Djakovitch, L.; Gloaguen, B.; Thiel, W.; Fillaut, J.; Delville, M.; Astruc, D.; Boese, R. *Angew. Chem. Int. Ed. Engl.* 1993, 32(7), 1075-1077.
[22] Valério, C.; Fillaut, J.; Ruiz, J.; Guittard, J.; Blais, J. C.; Astruc, D. J. *Am. Chem. Soc.* 1997, 119(10), 2588-2589.
[23] Cuadrado, I.; Casado, C. M.; Alonso, B.; Morán, M.; Losada, J.; Belsky, V. *J. Am. Chem. Soc.* 1997, 119(32), 7613-7614.
[24] Gruzdev, M.; Usolcseva, N.; Torre-Lorente, L.; Lattermann, G.; Domracheva, N. *Liquid crystals and their application* 2006, 4, 89-98.
[25] Gruzdev, M. S. Ivanovo: Dissertation; 2006.
[26] Webber, P. W.; Liu, Y. J.; Devadoss, C.; Bharathi, P.; Moore, J. S. *Adv. Mater.* 1996, 8(3), 237-241.
[27] Bosman, A. W.; Schenning, A. P. H.; Janssen, R. A. J.; Meijer, E. W. *Chem. Ber. Recueil.* 1997, 130(6), 725-728.
[28] Barefield, E. K.; Carrier, A. M.; Vanderveer, D. G. *Inorg. Chim. Acta.* 1980, 42, 271-275.
[29] House, D. A.; Robinson, W. T. *Inorg. Chim. Acta.* 1988, 141(2), 211-220.
[30] Vassilev, K.; Turmanova, S.; Dimitrova, M,; Boneva, S. *Eur. Polym. J.* 2009, 45(8), 2269-2278.
[31] Carvalho, N. M. F.; Horn, A.; Bortoluzzi, A. J.; Drago, V.; Antunes, O. A. C. *Inorg. Chim. Acta.* 2006, 359(1), 90-98.
[32] Nakajima, R.; Tsuruta, M.; Higuchi, M.; Yamamoto, K. *J. Am. Chem. Soc.* 2004, 126(6), 1630-1631.
[33] Thomas, K.; Velusamy, M.; Palaniandavar, M. Acta Crystallogr. Sec. C: *Cryst Struct Commun.* 1998, 54(6), 741-743.
[34] Richards, R.; Gregory, N. *J. Phys. Chem.* 1965, 69(1), 239-244.
[35] Wickman, H.; Klein, M.; Shirley, D. *J. Chem. Phys.* 1965, 42(6):2113.
[36] Aasa, R. *J. Chem. Phys.* 1970, 52(8), 3919-3930.
[37] Karlin, R. Magnetochemistry, *Mir:* Moscow, 1989.
[38] Domracheva, N.; Mirea, A.; Schwoerer, M.; Torre-Lorente, L.; Lattermann, G. *Chem. Phys. Chem.* 2005, 6(10), 110-119.
[39] Domracheva, N.; Mirea, A.; Schwoerer, M.; Torre-Lorente, L.; Lattermann, G. *Chem. Phys. Chem.* 2006, 7(12), 2567-2577.
[40] Menil, F. *J. Phys. Chem. Solids.* 1985, 46(7), 763-789.

[41] Higuchi, M.; Tsuruta, M.; Chiba, H.; Shiki, S.; Yamamoto, K. *J. Am. Chem. Soc.* 2003, 125(33), 9988-9997.
[42] Yamamoto, K. *J. Polym. Sci. Part A.* 2005, 43(17), 3719-3727.
[43] Domracheva, N. E.; Morozov, V. I.; Gruzdev, M.S.; Manapov, R. A.; Pyataev, A. V.; Lattermann, G. *Macromol. Chem. Phys.* 2010, 211(7), 791-800.
[44] Hadjipanayis, G. C., Prinz, G. A. (Eds.). Science and Technology of Nanostructured Materials; *Plenum Press:* New York, 1991.
[45] Pankhurst, Q. A.; Connolly, J.; Jones, S. K.; Dobson, J. *J. Phys. D: Appl. Phys.* 2003, 36(13), R167-R181.
[46] Dormann, J. L.; Fiorani, D. (Eds.). Magnetic properties of fine particles; North-Holland: Amsterdam, 1992.
[47] Esumi, K. *Top. Curr. Chem.* 2003, 227(Colloid Chemistry II), 31.
[48] Crooks, R. M.; Zhao, M.; Sum, L.; Chechik, V.; Yeung, L. K. *Acc. Chem. Res.* 2001, 34(3), 181-190.
[49] Cameron, J. H.; Facher, A.; Lattermann, G.; Diele, S. *Adv. Mater.* 1997, 9(5), 398-403.
[50] Facher, A. Bayreuth: Dissertation; 2000.
[51] Torre-Lorente, L. Bayreuth: Dissertation; 2007.
[52] Domracheva, N.; Mirea, A.; Schwoerer, M.; Torre-Lorente, L.; Lattermann, G. Fiz. *Tverd. Tela.* 2007, 49(7), 1392-1402.
[53] Floriano, P. N.; Noble, C. O.; Schoonmaker, J. M.; Poliakoff, E. D.; McCarly, R. L. *J. Am. Chem. Soc.* 2001, 123(43), 10545-10553.
[54] Choi, H. C.; Kim, W.; Wang, D.; Dai, H. *J. Phys. Chem. B.* 2002, 106(48),12361-12365.
[55] de Biasi, R. S.; Devezas, T. C. *J. Appl. Phys.* 1978, 49(4), 2466-2469.
[56] Raikher, Yu L.; Stepanov, V. I. *Phys. Rev. B.* 1994, 50(9), 6250-6259.
[57] Berger, R.; Kliava, J.; Bissey, J-C.; Baïetto, V. *J. Phys: Condens. Matter.* 1998, 10(38), 8559-8572.
[58] Berger, R.; Bissey, J-C.; Kliava, J.; Daubric, H.; Estournes, C. *J. Magn. Magn. Mater.* 2001, 234(3), 535-544.
[59] Nagata, K.; Ishihara, A. *J. Magn. Magn. Mater.* 1992, 104-107(3), 1571-1573.
[60] Gazeau, F.; Shilov, V.; Bacri, J. C.; Dubois, E.; Gendron, F.; Perzynski, R.; Raikher, YuL.; Stepanov, V. I. *J. Magn. Magn. Mater.* 1999, 202(2-3), 535-546.
[61] Gazeau, F.; Bacri, J. C.; Gendron, F.; Perzynski, R.; Raikher, YuL.; Stepanov, V. I.; Dubois, E. *J. Magn. Magn. Mater.* 1998, 186(1-2), 175-187.
[62] Berger, R.; Kliava, J.; Bissey. J-C.; Baïetto, V. *J. Appl. Phys.* 2000, 87(10), 7389-7396.
[63] Koksharov, YuA.; Gubin, S. P.; Kosobudsky, I. D.; Beltran, M.; Khodorkovsky, Y.; Tishin, A. M. *J. Appl. Phys.* 2000, 88(3), 1587-1592.
[64] Noginova, N.; Chen, F.; Weaver, T.; Giannelis, E. P.; Bourlinos, A. B.; Atsarkin, V. A. *J. Phys: Condens Matter.* 2007, 19(24), 246208.
[65] Ennas, G,; Musini, A.; Piccaluga, G.; Zedda, D.; Gatteschi, D.; Sangregorio, C.; Stanger, J. L.; Concas, G.; Spano, G. *Chem. Mater.* 1998, 10(2), 495-502.
[66] Dutta, P.; Manivannan, A.; Seehra, M. S.; Shah, N.; Huffman, G. P. *Phys. Rev. B.* 2004, 70, 174428.
[67] Schmidbauer, E.; Keller, R. J. *J. Magn. Magn. Mater.* 1996, 152(1-2), 99-108.
[68] Dormann, J. L.; D'Orazio, F.; Lucari, F.; Tronc, E.; Prene, P.; Jolivet, J. P.; Fiorani, D.; Cherkaoui, R.; Nogues, M. *Phys. Rev. B.* 1996, 53(21), 14291-14297.

[69] Zhang, L.; Papaefthymiou, G. C.; Ying, J. Y. *J. Appl. Phys.* 1997, 81(10), 6892-6901.

[70] Kuhn, L. T.; Bojesen, A.; Timmermann, L.; Fauth, K.; Goering, E.; Johnson, E.; Nielsen, M. M.; Mørup, S. J. *Magn. Magn. Mater.* 2004, 272-276(2), 1485-1486.

[71] Domracheva, N. E.; Pyataev, A. V.; Manapov, R. A.; Gruzdev, M. S. *Chem. Phys. Chem.* 2011, 12(16), 3009-3019.

[72] Maeda, Y.; Tsutsumi, N.; Takashima, Y. *Inorg. Chem.* 1984, 23(16), 2440-2447.

[73] Federer, W. D.; Hendrickson, D. N. *Inorg. Chem.* 1984, 23(24), 3861-3870.

[74] Federer, W. D.; Hendrickson, D. N. *Inorg. Chem.* 1984, 23(24), 3870-3877.

[75] Timken, M. D.; Strouse, C. E.; Soltis, S. M.; Daverio, S. A.; Hendrickson, D. N.; Abdel-Mawgoud, A. M.; Wilson, S. R. *J. Am. Chem. Soc.* 1986, 108(3), 395-402.

[76] Haddad, M. S.; Lynch, M. W.; Federer, W. D.; Hendrickson, D. N. *Inorg. Chem.* 1981, 20(1), 123-131.

[77] Haddad, M. S.; Federer, W. D.; Lynch , M. W.; Hendrickson D. N. *Inorg. Chem.* 1981, 20(1), 131-139.

[78] Timken, M. D.; Abdel-Mawgoud, A. M.; Hendrickson, D. N. *Inorg. Chem.* 1986, 25(2), 160-164.

[79] Timken, M. D.; Hendrickson, D. N.; Sinn, E. *Inorg. Chem.* 1985, 24(23), 3947-3955.

[80] Kannappan, R.; Tanasae, S.; Mutikainen, I.; Turpeinen, U.; Reedijk, J. *Polyhedron.* 2006, 25(7), 1646-1654.

[81] Sivasubramanian, V. K.; Ganesan, M.; Rajagopal, S.; Ramaraj, R. *J. Org. Chem.* 2002, 67(5), 1506-1514.

[82] Fujii. H.; Kurahashi, T.; Ogura, T. *J. Inorg. Biochem.* 2003, 96(1), 133.

[83] Canali, L.; Sherrington, D. C. *Chem. Soc. Rev.* 1999, 28(2), 85-93.

[84] Bottcher, A.; Birnbaum, E. R.; Day, M. W.; Gray, H. B.; Grinstaff, M. W.; Labinger, J. A. *J. Mol. Catal. A – Chem.* 1997, 117(1-3), 229-242.

[85] Bottcher, A.; Grinstaff, M. W.; Labinger, J. A.; Gray, H. B. *J. Mol. Catal. A – Chem.* 1996, 113(1-2), 191-200.

[86] Salmon, L.; Bousseksou, A.; Donnadieu, B.; Tuchagues, J. P. *Inorg. Chem.* 2005, 44(6), 1763-1773.

[87] Leovac, V. M.; Jevtovic, V. S.; Jovanovic, L. S.; Bogdanovic, G. A. *J. Serb. Chem.* Soc. 2005, 70(3), 393-422.

[88] Chervonova UV, Gruzdev MS, Kolker AM, Manin NG, Domracheva NE. *Russ J Gen Chem.* 2010; 80(10), 1954-1962.

[89] Gruzdev, M. S.; Chervonova, U. V.; Kolker, A. M.; Domracheva, N. E. *Rus. J. Str. Chem.* 2011, 52(1), 83-90.

[90] DeSimone, R. E. *J. Amer. Chem. Soc.* 1973, 95(19), 6238-6244.

[91] Bohan, T. L. *J. Magn. Res.* 1977, 26(1), 109-118.

[92] Richards, R.; Johnson, C. E.; Hill, H. A. O. *J. Chem. Phys.* 1970, 53(8), 3118.

[93] Domracheva, N. E.; Luchkina, S. A.; Ovchinnikov, I. V. *Rus. J. Coord. Chem.* 1995, 21(1), 26-32.

[94] Ovchinnikov, I. V.; Konstantinov, V. N. Radiospectroscopy of condensed matter; Editor: Zaripov, M. M., *Nauka*: Moscow, 1990, p.90.

[95] Konstantinov, V. N.; Ovchinnikov, I. V.; Domracheva, N. E. *Rus. J. Str. Chem.* 1984, 25(2), 186-193.

[96] Blomberg, M. R. A.; Siegbahn, P. E. M. *Theor. Chem. Acc.* 1997, 97(1-4), 72-80.

[97] Bleaney, B.; O'Brien, M. C. M. *Proc. Phys. Soc. B.* 1956, 69(12), 1216-1230.

[98] Griffith, J. S. The theory of transition metal ions; *Cambridge University Press:* Cambridge, 1961.

[99] Safo, M. K.; Gupta, G. P.; Watson, C. T.; Simonis, U.; Walker, F. A.; Scheidt, W. R. *J. Am. Chem. Soc.* 1992, 114(18), 7066-7075.

[100] Gutowsky, H. S.; Albert, S. *J. Chem. Phys.* 1973, 58(12), 5446-5452.

[101] Rao, P. S.; Subramanian, S. *Molec. Phys.* 1985, 54(2), 415-427.

[102] Ivanova, T. A.; Ovchinnikov, I. V.; Turanov, A. N. *Phys. Solid. State.* 2007, 49(11), 2132-2137.

[103] Dormann, J. L.; Fiorani, D.; Tronc, E. *Adv. Chem. Phys.* 1997, 98, 283-294.

[104] Davletbaeva, I. M.; Pyataev, A. V.; Kalachev, K. E.; Sadykov, E.K.; Manapov, R. A. *Polymer Science, Ser A.* 2006, 48, 612-617.

[105] Lang, G.; Marshall, W. *Proc. Phys. Soc.* 1966, 87(1), 3-35.

[106] Oosterhuis, W. T.; Lang, G. *Phys. Rev.* 1969, 178(2), 439-456.

[107] Domracheva, N.; Pyataev, A.; Manapov, R.; Gruzdev, M.; Chervonova, U.; Kolker, A. *Eur. J. Inorg, Chem.* 2011, 2011(8), 1219-1229.

[108] Gütlich, P.; Goodwin, H. A. (eds.) Spin Crossover in Transition Metal Compounds I, Topics in Current Chemistry; *Springer:* Heidelberg, 2004, p.233.

[109] Hayami, S.; Gu, Z.; Shiro, M.; Einaga, Y.; Fujishima, A.; Sato, O. *J. Amer. Chem. Soc.* 2000, 122(29), 7126-7127.

[110] Chervonova, U. V.; Gruzdev, M. S.; Kolker, A. M. *Russ. J. Gen. Chem.* 2011, 81(9), 1853-1858.

[111] Pretch, E. Evaluation of structure of organic compounds; *Mir:* Moscow, 2006.

[112] Becker, H.; Beckert, R.; Berger, W.; Domschke, G.; Fanghanel, E.; Faust, J.; Fischer, M.; Gentz, F.; Gewald, K.; Gluch, R.; Mayer, R.; Muller, K.; Pavel, D.; Schmidt, H.; Schollberg, K.; Schwetlick, K.; Seiler, E.; Zeppenfeld, G. *Organikum;* Berlin, 2008; v. I, p.504.

[113] Kazitsina, L. A.; Kupletskaya, N. B. Applications of the UV, IR, and NMR spectroscopy in organic chemistry; *Vysshaya Shkola:* Moscow, 1971.

[114] Nakanisi, K. IR spectra and structure of organic compounds; *Mir:* Moscow, 1965.

[115] Cheremisina, I. M. *Rus. J. Str. Chem.* 1978, 19(2), 286-300.

[116] Saleh, A. A.; Tawfik, A. M.; Abu-El-wafa, S. M.; Osman, H. F. *J. Coord. Chem.* 2004, 57(14), 1191-1204.

[117] Chervonova, U. V. Dissertation, Ivanovo, 2012.

[118] Chandra, S.; Kumar, U. *Spectrochimica Acta Part A.* 2005, 61, 219-224.

[119] Ismail, T. M. A. J. Coord. Chem. 2006, 59(3), 255-270.

[120] Tipugina, M. Yu.; Lomova, T. N.; Motorina, E. V. *Russ. J. Coord. Chem.* 2005, 31(5), 357-363.

[121] Chandra, S.; Gupta, L. K. *Spectrochimica Acta Part A.* 2005, 61(1-2), 269-275.

[122] Zaikin, V. G. Mass-spectrometry of synthetic polymers; *VMSO:* Moscow, 2009.

[123] Calvano, C. D.; Palmisano, F.; Zambonin, C. G. Rapid. Commun. *Mass. Spectrom.* 2005, 19(10), 1315-1320.

[124] Weber, B.; Carbonera, C.; Desplances, C.; Letard, J-F. *Eur. J. Inorg. Chem.* 2008, 2008(10), 1589-1598.

[125] Boussekou, A.; Nasser, J.; Linares, J.; Boukheddaden, K.; Varret, F. *J. Phys. I, France.* 1992, 2(7), 1381-1405.

[126] Spiering, H.; Kohlhaas, T.; Romstedt, H.;, Hauser, A.; Bruns-Yilmaz, C.; Kusz, J.; Gutlich, P. *Coord. Chem. Rev.* 1999, 190-192(1999), 629-647.
[127] Weber, B.; Jager, E-G. *Eur. J. Inorg. Chem.* 2009, 2009(4), 465-477.
[128] Weihe, H.; Güdel, H. U. *J. Amer. Chem. Soc.* 1997, 119(28), 6539-6543.
[129] Schafer, A.; Horn, H.; Ahlrichs, R. *J. Chem. Phys.* 1994, 100(8), 5829-5836.
[130] Collison, D.; Powell, A. K. *Inorg. Chem.* 1990, 29(23):4735-4746.
[131] Dingle, R.; Lines, M. E.; Holt, S. L. *Phys. Rev.* 1969, 187(2), 643-648.
[132] Carlin,. R. L.;. Burriel, R.; Rojo, J. A.; Palacio, F. *Inorg. Chem.* 1984, 23(15), 2213-2215.
[133] Bouslimani, N.; Clement, N.; Rogez, G.; Turek, P.; Bernard, M.; Dagorne, S.; Martel, D.; Cong, H. N.; Welter, R. *Inorg. Chem.* 2008, 47(17), 7623-7630.
[134] Owen, J. *J. Appl. Phys.* 1961, 32(3), 213S-217S.
[135] Okamura, M. Y.; Hoffman, B. M.. *J. Chem. Phys.* 1969, 51(7), 3128-3129.
[136] Ozarowski, A.; McGarvey, B. R.; Drake, J. E. *Inorg. Chem.* 1995, 34(22), 5558-5566.
[137] Wickman, H.; Klein, M. P.; Shirley, D. A. A. *Phys. Rev.* 1966, 152(1), 345-358.
[138] Aleksanderov, A. I.; Pashkova, T. V.; Barakhtenko, D. V.; Gruzdev, M. S.; Chervonova, U. V. *Liquid crystals and their application*. 2011, 4(38), 14-22.
[139] Domracheva, N. E.; Vorobeva, V. E.; Pyataev, A. V.; Manapov, R. A.; Zueva, E. M.; Gruzdev, M. S.; Chervonova, U. V. *Russ. J. Str. Chem.* 2013, 54(1 supplement), 16-27.

In: New Developments in Liquid Crystals Research
Editors: A. T. Petkoska and A. T. Broach
ISBN: 978-1-63482-456-9

Chapter 6

Alignment Technologies for Nematic Liquid Crystals

Anita Trajkovska-Broach*
University of Rochester (alumni), Rochester, NY, US

Abstract

An overview of alignment technologies for nematic liquid crystals, proposed and developed in the past two decades, is given. Rubbed polymer and obliquely-deposited inorganic films, as well as photo-irradiated, IB-bombarded and patterned surfaces, are briefly reviewed along with several novel alignment concepts. Reports related to the molecular origins of a preferred liquid crystal orientation on these surfaces are highlighted with a goal to address the long-lasting debate on the relative contribution of surface-liquid crystal molecular interactions and surface topography to the preferred mesogen orientation.

Keywords: Liquid crystal alignment, rubbed polymer films, obliquely-deposited inorganic films, photoalignment, ion beam irradiation, patterned surfaces, novel alignment concepts

1. Introduction

From its beginning with the invention of twisted-nematic liquid crystal display (LCD) concept in the early '70s, LCD industry has been growing rapidly into a multi-billion dollar industry today. LCDs are electro-optical devices containing liquid crystal (LC) layer, which undergoes changes in its optical characteristics under externally applied electric field. Apart from the first LCD applications in calculators and watches, devices based on LCD technology have been adopted in variety of applications, such as TV screens, computer and laptop monitors, video projectors, digital cameras, mobile phones, PDAs, etc. In most of these LCD-based devices, for successful operation, LC molecules should be appropriately aligned. LCD's visual changes occur exclusively due to the different orientations of LC molecules in

* Corresponding Author address; Email: trajkovska_anita@yahoo.com.

different states of device operation. There is so-called *field-free orientation* (LCD's OFF-state) and a range of predetermined distributions over the LC orientations under a range of applied voltages (LCD's ON-state). The field-free LC orientation is provided by the boundary conditions of a geometry confining the LC layer, which boundary conditions are dictated by the *alignment layers*. The alignment layer is usually coated on the inner side of both substrates, which comprise the LC layer in a typical LCD configuration [1-10].

In general, the basis for the LC molecular orientation on the surface of alignment layer is its anisotropy, physical and/or chemical, which yields an anisotropic arrangement of the adjacent LC molecules (mesogens). *Homogeneous (planar) alignment* occurs when the LC director aligns parallel to the alignment layer, whereas perpendicular LC alignment with respect to the surface is referred to as *homeotropic (vertical)*. In addition, the LC director might exhibit a certain polar angle with respect to the alignment layer surface, so-called pretilted alignment. The pretilt angle is an important parameter responsible for the uniform texture of LCD during operation and crucial in defining device performances, such as contrast, threshold voltage, and switching stability.

The scope of this chapter is to give an overview of LC alignment technologies and the evolving trends in this field in the past two decades. Emphasis will be put on the origins of LC molecular alignment on different types of alignment layers. In particular, rubbed polymer surfaces, obliquely-deposited inorganic layers, photoalignment, ion/plasma beam technologies, patterned surfaces, as well as novel LC alignment concepts, will be reviewed. The chapter does not cover all the excellent reports on LC alignment techniques, but with an objective to present the variety in this field, only selected examples will be presented. Without going into too much details, an extensive list of references is recommended for further reading.

2. Liquid Crystal Alignment Technologies

2.1. Rubbed Polymer Surfaces As Liquid Crystal Alignment Layers

Traditionally, uniform alignment of LC molecules is achieved by unidirectional mechanical rubbing (also, called buffing) of a polymer surface with a cloth, e.g., velvet. The first report on the alignment of LC molecules on rubbed surfaces dates back in 1911 [11], latter confirmed in 1944 [12], and it is still widely used in LCD industry. "Although used in cutting-edge technology, it is surprising that a vital part in the construction of displays, namely the alignment layer, relies essentially on a mechanical rubbing process, invented 100 years ago" noted Hoogboom, Raising, Nolte et al. [13].

Reasons for the wide use of rubbing are its advantages, such as it is straightforward, easy and inexpensive way to orient mesogens with low capital investments. The resulting rubbed polymer alignment surfaces can provide stable and strong anchoring of LC molecules. Some of the polymers used in the rubbing method can provide also the pretilt angle in LC orientation, which is a crucial factor, responsible for uniform device texture during operation.

As polymer surfaces, variety of polyimides (PIs), Nylon, polyvinyl alcohol (PVA), polystyrene (PS), teflon-like polymers and others, have been explored and used as alignment layer materials [14-23]. Among them, PIs have been and are still continuing to be the most used materials for LC alignment due to their outstanding photo-, chemical and thermal-stability. Particualrly, PIs are considered high performance materials with high glass transition temperature, excellent dimensional stability, as well as good mechanical and dielectric properties, optical transparency and adhesion to variety of substrates. When used as LC alignment layers, PI films are known to provide stable anchoring of LC molecules on the film surface, which is critical for reliable LCD devices. Moreover, the diversity in PI chemical structures offers many possibilities for molecular interactions with the overlaying LC molecules, and thus, a spectrum of LC orientations, ranging from homogenous to homeotropic. Many parameters, such as chemical structure, molecular weight, groove size/groove strength, crystallinity and so on, have been studed in relation to LC alignment [15-18, 24-29].

The contact between the rubbing cloth and the polymer layer during the rubbing process creates microscopic-size grooves on the surface (surface "scratches", Figure 1a), but also preferentially orients the individual polymer backbones or their segments in the alignment layer. The preferential molecular orientation at the polymer surface has been determined to be parallel to the rubbing direction and it is believed to be occurring due to the pulling action of the rubbing cloth fibers [13]. The LC molecules in contact with such surface adopt the preferential surface orientation, and then, due to the cooperative nature of LC molecules, the surface LC orientation progresses into the bulk of LC material (Figure 1b).

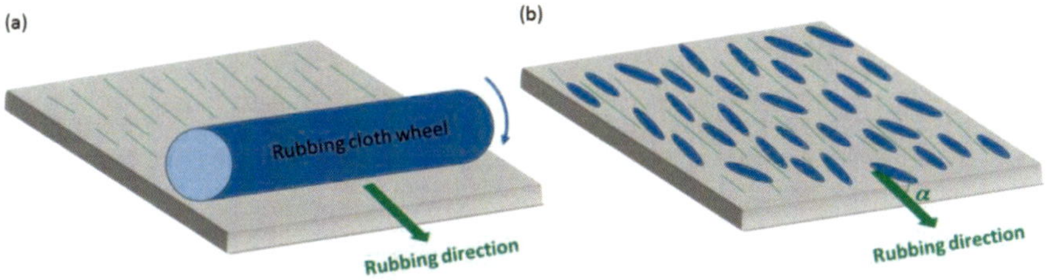

Figure 1. Rubbing of polymer surfaces. (a) Surface "scratches are formed by the rubbing cloth fibers. (b) LC molecules orient on rubbed surfaces usually in a direction parallel to the rubbing direction and form a pretlt angle α.

The origin of LC molecular alignment on alignment layer surfaces, especially those which involve anisotropic topography, *viz.* rubbed polymer layers, has been a subject for a debate for decades [30-37]. The proposed alignment mechanisms, particualrly associated with the rubbing method, can be classified into two categories. The first mechanism leading to orientation of LC molecules on rubbed surfaces is the *"physical" mechanism*, i.e. the formation of microgrooves on the surface is solely responsible for LC alignment (Figure 2a). The basis for this purely physical mechanism of alignment was initially described by Berreman, in his elastic continuum theory [38]. The molecular interactions between the alignment layer and the LC molecules are not taken into account and the LC molecules orient usually parallel to the microgrooves formed during the rubbing (Figure 2a). LC molecular orientation parallel in and around the microgrooves results in the lowest surface energy, and hence, the most stable conformation. The originally proposed Berreman's model has been shown that cannot explain all the cases of surface-groove-induced anchoring of a nematic LC molecules and it has been re-examined [39,40]. The second mechanism of the origin of LC alignment on rubbed surfaces is the *"molecular" mechanism,* i.e. the interactions between the

LC molecules and the alignment layer surface are responsible for the alignment [16-18,35,41]. The basis for this mechanism are the numerous findings that the individual polymer chains and/or their segments and substituents are preferentially oriented on the surface by the unidirectional rubbing method, which then engage in anisotropic molecular interactions with the overlaying LC molecules.

Earlier reports on the origins of LC alignment claimed the "need" of anisotropic topographic features for efficient LC orientation ("physical" model). With the development of not only rubbed surfaces, but also other types of LC alignment layers, which have solely chemical anisotropic surface and not any anisotropic physical surface features, the scientific evidence is in favor of the anisotropic surface-LC molecular interactions as main reasons for LC orientation on such surfaces ("molecular" model). However, the nature of the LC-surface interactions is complex, and includes but not limited to, van der Waals interactions, dipole-dipole interactions, π-π interactions, steric factors, chemical and/or physical (hydrogen) bonds. Moreover, the relative contribution of intermolecular interactions and topography (if present) to a preferred LC orientation, is not clear.

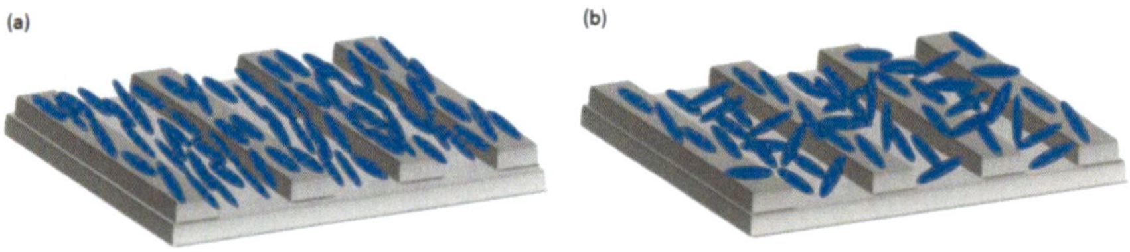

Figure 2. LC molecules on microgrooved surfaces. (a) Berreman "physical" model. (b) Possible LC orientations on surfaces with micron-size grooves.

In the text below, several reports addressing the origins of mesogen orientation on rubbed surfaces will be briefly discussed. Stohr et al. proposed a microscopic bond orientation model for potential origins of LC alignment on rubbed polymer surfaces based on experimental evidence. Particualrly, surface-sensitive and polarization-dependent near-edge X-ray absorption fine structure (NEXAFS) measurements revealed a preferred in-plane and out-of-plane orientation of phenyl and C=O groups at the surface of rubbed PI films. This unidirectional molecular alignment at the surface is responsible for LC alignment on these surfaces. Both, the LC orientation along the rubbing direction and the direction of the out-of-plane LC pretilt can be explained with this model. LC direction follows the preferential orientation of the phenyl rings at the surface. The proposed LC alignment model does not require the existence of crystalline order as some of other models predict [31,32].

Using semi-empirical molecular orbital methods, Koenig et al. have calculated the local dipole moment of the rubbed polymer films and found a direct correlation between the electro-optical behavior of the LC cell and the dipole moment for several polymeric alignment layers used in those cells. This is a direct evidence that the LC orientation on the surface is determined by interactions at the LC/alignment layer interface, which in this particular study, are dominated by the dipole–dipole forces [35].

Furthermore, series of rubbed well-defined brush-PIs have been studied by Ree et al. [17,18]. The microgrooves generated on the polymer surface by the rubbing process are different in size, mostly in micron and submicron range, which is much larger than the mesogen size; therefore, the LC molecules can align in any direction inside the micron-sized surface groove (Figure 2b). From the other hand, the polymer chain segments in the alignment layer are well matched with the LCs in size, and thus, can more effectively interact with the LC molecules, compared to the microgrooves, which are mismatched significantly. Therefore, the observed LC alignment and relatively high anchoring energy suggest that the LC molecules are more likely to be aligned by favorable anisotropic interactions with the oriented polymer main chains and the alkyl end groups in the bristles, rather than by the microgrooves. Further, the authors even speculate of potential interactions, *viz.* the aromatic mesogen might favorably interact with the phenyl components of the PI backbone via *π-π interactions*, the polar cyano end in the mesogen interacts with the imide rings in the backbone via *dipole-dipole interactions*, while the nonpolar alkyl tail interacts with the alkyl end groups in the bristles via *van der Waals type interactions*. In another study on rubbed brush-PI films of various lengths of the bristles, the same group observed perpendicular and parallel LC alignment behaviors depending on the length of the bristles. Besides the microgroove formation and orientation of the polymer backbones along the rubbing direction, the alkyl bristle end groups were found to be oriented along the rubbing direction, but the phenyloxy bristle units were found to be aligned perpendicular to the rubbing direction. The resulting LC alignment was suggested to be *a play-off between the anisotropic interactions* of LC molecules with all the possible (parallel and perpendicular) oriented segments at the film surface [18].

In another report on rubbed PS surfaces, surface analysis showed the perpendicular orientation of the phenyl rings with respect to the rubbing direction. Despite the reports on existence on perpendicular sub-micron grooves in addition to the main microgrooves on the rubbed on PS surface, it is believed that the mesogens' interactions with the perpendicularly reoriented phenyl side groups during the rubbing overcome the effects of the parallel reoriented vinyl main chains by the rubbing and are responsible for LC alignment [19,20].

When the rubbed PI film surface was covered with thin metal layer, the surface lost its LC alignment capability leading to the conclusion that the LC alignment is governed by the oriented backbones, and consequently, the anisotropic interactions with the overlaying mesogens [19].

The origin of another important parameter, the pretilt angle α, found on some rubbed surfaces, is also not well-understood. Not only the tilt direction, but also the magnitude of the pretilt angle is important. For rubbed PI films, it has been found that the tilt direction usually points up into the rubbing direction, as shown in Figure 1b, while the magnitude of the overlaying LC pretilt angle α was found to be related to the tilt angle of the aromatic building blocks at the PI surface. Attempts to design mateials that will yield a controllable LC pretilt have been made. For instance, PI-based material structures with bulky side groups and long alkyl chain groups enabling vertical alignment, soluble PI structures and others, have been proposed to provide more controllable tilted LC orientation and "easy" processing [24-29,42].

PIs with different inclination angles of polymer backbones have been developed and the tilt angle of the aromatic building blocks in the PI backbone have been linked the generated LC pretilt angle [16].

Stable vertical alignment of nematic LC was observed on PI-based polymers and copolymers with mesogenic side chains (pretilt angles 89-90^{0}) before and after rubbing. This study concluded that the formed micro-grooves caused by the rubbing do not affect the LC alignment, which is mainly determined by the conformation of the side mesogenic groups and their strong interactions with LC molecules [24-26].

In another study, surface morphology of the rubbed PI film was shown to be strongly influenced by the PI backbone conformation, as expected. The interactions between LC molecules and PI alignment layer have been shown to have a greater effect on LC pretilt angles than the film surface morphology. No correlation between the LC pretilt angle and the surface energy of PI film was found. However, side groups attached to PI backbone, especially nonpolar or bulky ones have significant effects on the pretilt angle. For instance, PIs made with both nonpolar fluorinated groups sticking out of the surface and the polar hydroxyl groups on the surface exhibits high pretilt angle [16].

A way to generate a wide spectrum of pretilt angles was demonstated by using a rubbed-treated PI layer consisting of a mixture of homogeneous PI and homeotropic PI materials. Uniform LC alignments with various pretilt angles in the range from 0-60° were measured, where the pretilt angle was expressed as a function of the concentration of homeotropic PI [43]. Another study reported stacking of a vertical PI alignment layer on top of a planar PI alignment layer, and in such way, accomplished controlled pretilt angles ranging from 0° to 90°depending on the thickness of the vertical alignment layer [44]. Yet another study was using PI blend films prepared from two types of polyamic acids, one for vertical LC alignment and the other for planar LC alignment. The study showed a good control over the pretilt angles by variations in the baking process. The resulting LC alignment behavior was correlated with the wettability of the PI films, which varied due to the fragmentation of the long alkyl side group on the PI surfaces by the baking process [45].

Realizing the importance of molecular interactions, trials to modify them via nanoscale hybridization, have been shown as a potential way to tune the alignment properties of rubbed PI layers. Particularly, by doping or physical dispersion of polyhedral oligomeric silsesquioxane (POSS) nanoparticles in homogenous commercially available PIs, the surface energy of the PI alignment films was found to decrease, and at the same time, a controllable pretilt in the range 0 – 90°, can be generated, which cannot be easily achieved by complicated synthesis of tailored PIs [46-48].

Melallic surfaces have been explored for LC alignment and upon rubbing yielded planar orientation [23].

Despite its broad use and the efforts directed toward development of the associated materials (PIs, for example), the rubbing method has some serious drawbacks, which are the result of the direct contact between the alignment layer and the velvet cloth. The rubbing creates debris and electrostatic charge, which can interfere with the uniform LC alignment and display pixel switching, resulting in pixel damage, and thus, lower production yields and higher costs due to unacceptable deterioration in display performance. Moreover, as parts of polymer layer are rubbed-off, the contrast ratio can vary from one surface site to another. Besides the issue with the associated debris, the rubbing process is done in clean room and is not applicable on flexible substrates. In addition, multi-domain LC alignment needed for e.g., wide viewing angle LCDs is hard to achieve with conventional rubbing [13].

To overcome the shortcomings of the rubbing process, several non-contact (rubbing-free) alignment methods have been proposed. Among them, obliquely-deposited inorganic layers, ion beam bombardment, photoalignment approach and patterned surfaces have shown promise in replacement of the rubbing method.

2.2. Obliquely-Deposited Inorganic Liquid Crystal Alignment Layers

Inorganic thin films have been widely explored as LC as alternative alignment layers to the rubbing method, but also for applications where high thermal, photo and chemical stability is important, e.g., in liquid-crystal-on-silicon (LCoS) applications.

Among the inorganic materials, mostly films of SiOx, Al_2O_3, TiO_2, ITO, ZnO, ZrO_2 and other metal oxides have been proposed. Various deposition techniques, *viz.* RF sputtering deposition, reactive sputtering, or electron beam evaporation, usually at oblique or glancing angles, have been utilized for fabrication of these layers (Figure 3) [49-63].

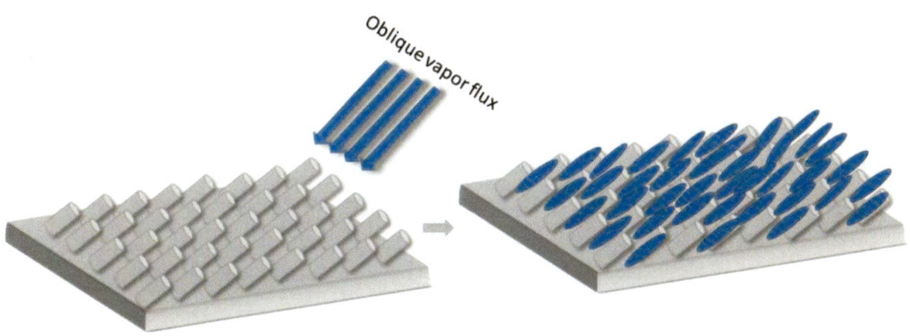

Figure 3. Deposition of metal oxides under oblique angle results in anisotropic topography, which is sufficient for a preferential LC orientation.

The relationship between the RF-magnetron sputtering conditions and the LC alignment properties of SiOx thin films was studied. The behavior of LC molecules on top of these films was mainly affected by the surface energy of SiOx thin films, which was closely related to the chemical composition of SiOx. Particularly, the azimuthal anchoring energy and the pretilt angle of LC molecules were found to vary by the oxygen content x in SiOx [49]. A vertical and homogenous LC alignment has been achieved on a-SiOx thin films. LC molecules aligned vertically on a-SiOx film when the stoichiometry parameter x was smaller than 1.56, but they aligned homogeneously for x larger than 1.56 [50].

A columnar topography was reported in obliquely deposited SiO_2 alignment films deposited by electron beam evaporation [56]. Anchoring transitions were observed for nematic LC on obliquely evaporated SiOx. The LC anchoring on such SiOx was speculated to have two origins: long-range van der Waals and short-range surface interactions; each with opposite preference in the LC director orientation, and thus, the observed transition was found to be dependent on their relative strength with the overlaying mesogens [51].

The vertical alignment of LCs having negative dielectric anisotropy on an amorphous SiOx (a-SiOx) thin film is due to the anisotropic interactions between LC and a-SiOx film. To investigate the mechanism of vertical alignment, physico-chemical characteristics of the alignment layers were varied by controlling the composition. Particulalry, the variations in a-SiOx composition yielded changes in the polarizability, which is a simple measure of the induced-dipole strength at the surface of the alignment layer. There is a critical transition point from planar to vertical alignment of the overlaying LC, and it was found that the long-range van der Waals interactions is responsible for the vertical alignment. Furthermore, the competition between long-range van der Waals interaction and short-range dipolar interaction was studied in terms of the interfacial energy between mesogens and an alignment layer [55].

Although, inorganic alignment films have been shown to be capable of generating the desired LC orientation, the fabrication of inorganic alignment layers is intrinsically expensive and not very compatible with continuous production processes.

2.3. Photoalignment of Liquid Crystals

The photoalignment has been proposed as "clean" rubbing-free method offering many advantages, such as the absence of contamination associated with rubbing method, feasibility of multi-domain pixel generation, low temperature processing, and thus, application on flexible substrates.

Photoalignment is a non-contact alignment method, which utilizes linearly-polarized (LP) or unpolarized light to generate chemical anisotropy on the surface of photo-sensitive layer (Figure 4). The method actually relies on anisotropic photoreaction or directional light-induced reorientation of certain chemical species on the surface of alignment layer, which consequently yields a surface with chemical anisotropy, sufficient to orient the overlaying mesogens mostly by anisotropic molecular interactions [1,8,10,64-84,91,142,159,160].

The first report on photoalignment dates back in 1988 [64], when Ichimura et al. showed that the trans/cis photoisomerization of azo-benzene monolayer, also called a *"command surface"*, can switch the alignment of the overlaying LC molecules between homeotropic and planar modes. Soon after, Gibbons et al. doped the azo dye in a a standard PI lalignment layer [65-67], while Schadt et al. [68-71] showed that angular selective excitation by LP light causing directional surface photo-reaction (particularly, photo-crosslinking) is sufficient to provide alignment of LC molecules in contact with the exposed surfaces. Irradiation of widely used PIs have been also explored for LC alignment [72,73]. These initial reports prompted an intensive research in the field of photo-sensitive materials and processes using light for their potential as LC alignment "tools" [74-169]. Excellent books and reviews on photoalignment technology have been published by Hasegawa, Chigrinov, Ichimura, Seki, Kelly and Reznikov, among the others [1,10,74-77].

According to the photochemical reaction and/or photo-induced reorientation responsible for the LC alignment, three main classes of materials can be distinquished, as follows:

(i) azo-containing compounds undergoing reversible *cis–trans* isomerization or photo-induced diffusion reorientation,
(ii) compounds containing cinnamoyl, coumarin, or chalcone photo-reactive groups, undergoing directional photochemical cross-linking, and
(iii) PI-based compounds undergoing photo-degradation and other orientational bond breaking.

Combinations of photo-induced reaction, isomerization and/or diffusion reorientation have been explored, as well [96,129,130,162-164,167].

2.3.1. Photo-Induced Cis-Trans Isomerization and Reorientation of Azo-Compounds

A variety of azo-containing material structures that utilize cis–trans photoisomerization or photo-reorientation have been developed. By the absorption of polarized light, the azo dye undergoes isomerization or diffusion rearrangement, which affects the spatial arrangement of adjacent LC molecules. The azo-dye can be doped directly into a conventional aligmment layer, e.g., PI, or covalently bonded to a polymer in the main chain or as a side group, or can be a part of Langmuir–Blodgett (LB) layers or self-assembled monolayer (SAM) or polymer network compositions [64-67,74,75,78-98].

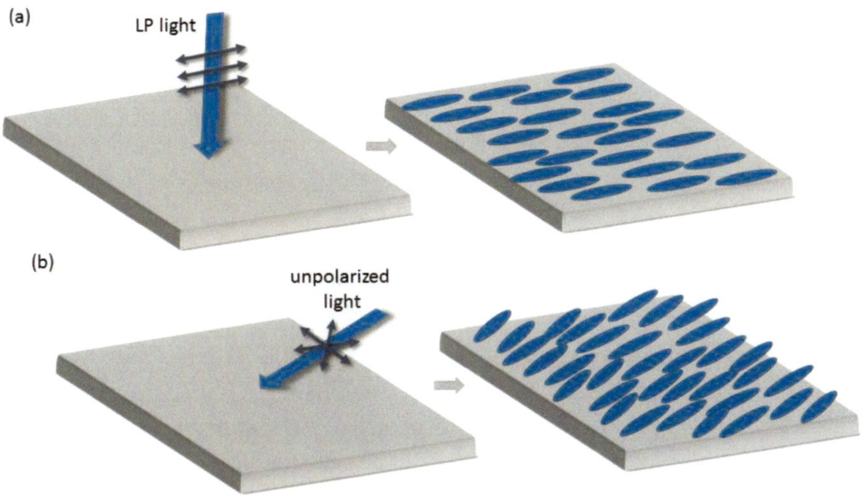

Figure 4. Basics of photoalignment. (a) Exposure of photosensitive surface with polarized light at notmal incidence resulting in planar LC alignment. (b) Exposure of photosensitive surface with unpolarized light at oblique incidence yielding tilted LC orientation.

Three-dimensional manipulation of LC molecules by light have been demonstrated on azo-containing surfaces by changing the incident direction of the actinic unpolarized light: the net-orientation of azobenzene molecules, and thus, the surrounding mesogens were oriented along the propagation direction of the actinic light.

The initially proposed structures of azo compounds required the use of intense light and/or large doses of light, to induce the isomerization. Also, high concentration of azo-dye was needed in order to have an efficient effect on the surrounding mesogens, which can cause an undesirable color of the alignment layer. The thermal and UV stability of the alignment layers with azo-dyes were insufficient due to relaxation effects or degradation. However, over the past decade, more stable and more light-sensitive azo-compounds have been proposed. For instance, Chigrinov, Kwok et al. introduced the sulfonic-azo dye derivatives capable of the so-called *photo-induced diffusion reorientation*, as a different way of causing photo-induced changes without involving any photochemical reaction or structural transformations of the molecules. The resultant photoaligning films needed very low light dose (less than 50mJ/cm^2) and were found to be photochemically stable and robust with excellent LC alignment ability enabling a controllable pretilt angle and strong anchoring energy. Besides different stable sulfonic-azo dye structures, the same group has also proposed sulfonic azo dye monomers, which upon photo-induced alignment, can be polymerized [84,87-90]. Other authors have confirmed the photo-induced reorientation or photo-ordering of azo compounds [93,94].

Transitions between different LC alignment modes have been proposed by photo-active azo-surfaces due to the excellent reversibility of trans–cis and cis–trans isomerization.

Gibbons et al. have demonstrated a PI film doped with azobenzene molecules as a dichroic dye and found that the homogeneous alignment of LC molecules can be altered by LP light exposure of the doped PI film [65-67]. Later, other research groups have demonstrated reversible switching in LC orientation between two in-plane LC orientations, or planar-vertical LC switching alignment, or even between ordered and disordered LC state [64,78,95].

In this context, azo-containing SAMs have been widely explored for LC alignment in terms of their molecular structure, packing density, etc. One study has shown that azo-silane SAMs with long chains provided a stable homeotropic alignment even after photobuffing, while azo-silane SAMs with short chains showed photo-switchable homeotropic and planar alignments. However, when the packing density was increased to a certain degree, even azo-silane SAM layers exhibited stable homeotropic alignment [169].

2.3.2. Anisotropic Photo Cross-Linking

Photoinduced anisotropic cross-linking (particularly, photo-dimerization) of materials containing photo-reactive groups, such as cinnamate, coumarin or chalcone, seems the most promising sub-category of photo-induced LC alignment [68-71,103-119]. Among the first materials studied in this category was poly(vinylcinnamate) (PVCi), which films upon exposure to LP UV light yielded homogeneous LC alignment perpendicular to the polarization direction of the incident light. The alignment was the result of mostly anisotropic cross-linking of the cinnamate groups via (2+2) cycloaddition reaction, although cis–trans isomerization can occur in cinnamates at low fluences. However, PVCi and most of the initially proposed cinammate-based structures suffered from insufficient thermal- and photo-stability, as well as low pretilt and limited control of LC pretilt angles [76]. More stable LC alignment was observed in some of the other proposed cinnamate-based and coumarin-based structures [69-71,101-110].

Attempts to stabilize the pretilt angle in cinnamate-containing photoalignment materials have been made. A simple method to generate the pretilt angle, comparable to that achieved by the rubbing method, was proposed by a simple directional peel-off process using a photo-reactive cinnamate oligomers with flexible molecular chains [102].

To stabilize the pretilt angle in PVCi films, Reznikov et al. use LP UV exposure in the presence of a magnetic field. Large and stable pretilt angles in the overlaying LC layer were measured contrary to the LP UV irradiation in the absence of a field, when a rapid decay of the pretilt angle was observed [120,121].

Ichimura et al. suggested that highly sensitive photoalignment of LCs can be realized by axis-selective triplet energy transfer. Addition of a triplet photosensitizer into a photocrosslinkable polymer with cinnamate side chains yielded significant enhancement of its photosensitivity with very low exposure energies needed. This procedure can be very advantageous for high-throughput fabrication of LC-based devices involving LC photoalignment [118].

The pioneering work of Schadt et al. of inducing directional photoreaction of prepolymers by polarized UV light, also termed as *linear photo-polymerization (LPP),* is worth-mentioning for his enormous contribution to the field of LC photoalignment, and in particular, for establishing the foundation of cutting-edge photoalignment materials and processes developed by Rolic Technologies [68-71]. With the help of LPP technology, Schadt and his team have demonstrated different optical contact-free micro-patterned LC orientations leading to high resolution wide viewing angle multi-domain LC alignment in several LCD modes, as well as in number of novel passive optical elements (compensators, security elements, etc.). They have also proved that pretilt angles can be generated under oblique exposure of photoreactive materials. For the LC pretilt angle on cross-linkable surfaces, they found that it is "easily" achievable by exposure at oblique incidence, when the LC alignment direction is parallel to the polarization direction of the incident UV light, like in coumarin-containing polymers. The anchoring energy of coumarin-based compounds was found to be comparable to rubbed PI [70].

Chen et al. looked into the origins of LC alignment on various polymethacrylates with coumarin-containing side chains. Thin films of these materials, when irradiated with LP UV light at normal incidence, undergo dimerization process and can serve as efficient LC alignment layers. The attempts to elucidate the roles played by unreacted coumarin monomers and coumarin dimers were made. Particularly, three factors, such as: (a) the relative abundance of coumarin monomers and dimers, (b) their evolving orientational orders, and (c) the energetics of their interactions with the overlying LCs, were studied experimentally and theoretically. A kinetic model was developed to evaluate the relative importance of these three factors on LC orientation and explain the observed cross-over behavior in LC alignment with LP UV exposure from initial parallel LC orientation to the polarization plane of the incident polarized UV light to perpendicular LC orientation occurring at higher UV fluences [103-110].

Kawatsuki et al. have even developed photo-reactive liquid crystalline polymers (LCPs) with biphenyl group in the side chains as alignment layers for LCs. The initial alignment direction was parallel to the polarization plane of the LP-UV light at lower doses of irradiation and it changed to perpendicular alignment with increase in the exposure dose, as observed in other materials. The LC alignment direction was controlled mainly by the direction of the biphenyl group, inserted in the photo-crosslinked side groups, as well as by the degree of photo-reaction [111-116].

A polysiloxane-cinnamate based materials were also irradiated with LP UV light. Under short exposure times, the orienting surface provided an orientation of LC molecules parallel to the direction of UV light polarization, while higher exposure doses led to the orthogonal to the UV polarization orientation of LC. In addition, it was observed a range of exposures that give a bistable anchoring conditions and domains with different directions of the director orientation. The authors proposed a phenomenological model of this anchoring transition and surface bistability based on the competition between: (a) trans-cis isomerization of the side fragments of the polysiloxane-cinnamate material, and (b) cross-linking photoreaction of side fragments resulting in orthogonal molecular easy axes [145-148,162].

2.3.3. Anisotropic Photo-Degradation of Polyimides

Various PI molecular structures have been extensively studied for their potential use as photoalignment materials, mostly due to their already wide use in LCDs as rubbed alignment layers, which are able to provide any LC alignment, homogenous, tilted and homeotropic. Before the UV exposure, the PI chains are randomly oriented in the layer. During the irradiation, LP UV light excitates selectively the chains whose long axis has similar or identical direction as the polarisation direction of LP UV light. The direction perpendicular to the polarisation direction of incident UV light will be the direction with the highest density of unbroken PI chains. The LC molecules will be oriented on such layers, mostly in perpendicular direction with respect the the polarization plane of LP UV, due to the van der

Waals interactions with "unaffected" chains by the UV light. Furthermore, the formation of pretilt angle by oblique unpolarized light exposure is possible due to the homogeneous LC orientation in the direction corresponding to the maximum density of unbroken polymer chains. However, besides the radical formation, if the exposure is done in air atmosphere, subsequent oxidation reactions might occur in PI films, and thus, generation of a "chemical debris". The anisotropic photo-degradation of PIs usually requires more energy than the other photochemical materials and processes used for LC photoalignment [72,73,76,122,127, 131,161,165].

Selected examples of modified PI structures yielding more controllable LC alignment and pretilt are given below. Unidirectional LC alignment by LP UV exposure of PI with a fluorene unit incorporated in a side chain showed LC alignment parallel to the exposure polarization of UV. Pretilt angles in a range 0° to 90° were obtained in fluorine-containing PI films. Depending on the fluorine content and the UV exposure time, which is closely related to the surface energy of the photoalignment layer, the pretilt angle can be controlled [128].

In situ exposure to LP UV light during thermal imidization of poly(amic acid)s has been found to result in a parallel LC alignment to the direction of polarization of LP light, whereas the conventional method, in which the alignment layers are exposed to LP light after imidization, leads to alignment perpendicular to the direction of polarization of the LPL [159].

To avoid the formation of free radicals and other "chemical debris" in the photoalignment films, a wide spectrum of PI structures containing photoreactive (photo-crosslinking) groups in the backbone or as side groups have been developed [123,125,163,164].

An inclined homeotropic alignment of LC molecules was achieved by using photo-sensitive films of a PI with azobenzene in the backbone. Double light exposure method was applied: first with LP light exposure at normal incidence followed by the oblique angle irradiation with unpolarized lights. The LC pretilt angle was controlled by the second step - oblique angle irradiation with unpolarized light. In another study, a method of repeated cis-trans photoisomerization reaction resulting in the reorientation of a backbone structure of polyamic acid containing azobenzene units was reported. The subsequent imidization seems to stabilize the azo-PI structure. The LC alignment was thermally and optically stable and no decomposition of PI backbone was involved.

PI-based structures with a pendant cinnamate group were reported, as well. The LC alignment on such films exposed to LP UV light was independent on the cinnamate content, but the thermal stability of the photo-induced LC alignment was significantly improved by reducing cinnamate content. This could be attributed to the strong interchain interaction of PI chains, which prevents thermal randomization of the photo-products of pendant cinnamates.

PI films with chalcone derivatives using LP UV light were studied, where the LP UV light irradiation generated dimerization and isomerization products of the chalcones. The alignment directions of LC molecules were found to be either homeotropic or planar depending on the alkyl chain length in the polymer structure [163].

Attempts to stabilize PVCi films were also made by blending PVCi and PI, among the other approaches. The LC alignment on PVCi/PI blend alignment films sustained after thermal treatment at 200°C, which has been attributed to the fixation of photo-dimer of cinnamate by the rigid PI chains in the blend alignment layers. To further modify the surface properties, fluorinated PI blends with PVCi were prepared, which gave rise to the pretilt angle of LC with increasing the fluorine content due to its role in lowering surface energy [129].

Materials having dual photoreactive side moieties have been proposed for better control of the pretilt angle. Particulalry, PI containing cinnamoyl and coumarin side groups were synthesized in a high molecular weight. Upon LP UV exposure, the both photosensitive side groups undergo (2+2) photodimerization, causing reorientation of the polymer chains, including the polymer backbone, in a preferential direction. A series of PIs containing main-chain chalcone groups and side-chain cinnamate or acetate groups were synthesized. The alignment ability and pretilt angles were found to be dependent on the structure and exposure level. Novel photoreactive PI structures with cinnamoyl, acryloyl and methacryloyl side groups, as well as photosensitive PI structures with stilbene in the backbone were synthesized [130]. PS derivatives with azo-benzene and cinnamate-groups yielded planar LC alignment with pretilt angles in the range 0.5 - 5°, which was dependent on the molar contents of cinnamate and azo-benzene groups. The irradiation with LP-UV light lead to directional cross-linking the cinnamate groups, followed by the irradiation of linearly polarized visible (LP-Vis) light, which induced isomerization of azo-benzene groups.

Attempts to expose surfaces via photomask yielded patterned topographies in addition to anisotorpic surface photo-reactions. For instance, LP UV laser light exposure of a photo-sensitive PI film via mask resulted in grooves formed in the exposed regions, but they had little or no effect at all on LC orientation. The microembossed structures (microgrooves) of the polymer films were generated by (2+2) cycloaddition reaction of the photosensitive cinnamoyl side group by the exposure of ultraviolet (UV) light through photomasks [137]. Photoalignment coupled with surface relief gratings have been explored, as well [138,139].

Combinations of photoalignment and rubbing method, as well as ion beam exposure and photoreaction have resulted in good LC alignment [132-136].

Transition between different LC orientations on PI-based photosensitive surfaces, similar to azo-containing "command surfaces", have been demonstrated. For instance, Komitov et al. reported transition from vertical to planar alignment on PI alignment layer exposed to LP UV light, depending on whether the alignment layer was mechanically rubbed or not prior the illumination. The PI material was designed originally to promote vertical alignment. The mechanism behind this transition was associated to the inclination of the alkyl side chains of the PI alignment layer towards the rubbing direction and the bond breaking process due to UV illumination which are considered as the key factors for occurrence of such alignment transition [155,156].

Despite the advantages that photoalignment is offering over the rubbing method, the LCD industry is still heavily based on rubbed polymer alignment layers. Recently, after almost three decades of intensive research in the field of LC photoalignment, Sharp has adopted a photoalignment technology, originally developed by Rolic Technologies, in the fabrication of advanced new generation LCD-TV flat panel displays with reduced production costs and low energy consumption. The proprietary Rolic® Light Controlled Molecular Orientation (LCMO) [141], particularly LCMO-VA technology for vertical alignment LCDs is foundation for Sharp's Ultra-violet induced multi-domain vertical alignment (also called, UV^{2}A) technology. UV^{2}A technology offers highly accurate control over the LC alignment in accordance with the direction of the radiation. The photoalignment material, originally designed to provide homeotropic LC alignment, upon UV exposure gives slight inclination with respect to the surface normal. The simple process for production, successfully implemented in Gen 10 factory, and very precise control of LC modular orientation enables "real black state" and seemingly brighter screens. According to Sharp, there are numerous advantages of the panels made with UV^2A technology, like 20% higher aperture ratio, 60% higher contrast, and two times faster response time than the conventional panels [140].

2.4. Liquid Crystal Alignment on Patterned Surfaces

Various lithographic methods, such as photolithography, soft lithography, and nano-imprint lithography (NIL) have been utilized to pattern surfaces to "mimic" the rubbed surfaces, but in a more controllable and reproducible way. Direct laser writing, e-beam exposure and AFM tip nano-rubbing have been also proposed for making patterned surfaces suitable for LC alignment [170-193].

Surface grooves with a suitable pitch and depth can be very effective in aligning LCs. as mentioned above, the surface "scratches" caused by rubbing method, are directional but non-uniform, ranging from micron to sub-micron sizes, which are much larger in size than the LC molecules. From the other hand, soft and nanoimprint lithography can generate grooves and other periodic features with uniform and precise dimensions leading to more reliable LC alignment, which is hardly possible with the conventional rubbing process. Lithogrpahy is actually capable to create nanometer-resolution two-dimensional surface grooves on substrates over a large area by a step and stamp process (Figure 5). Materials to suit this process have been developed.

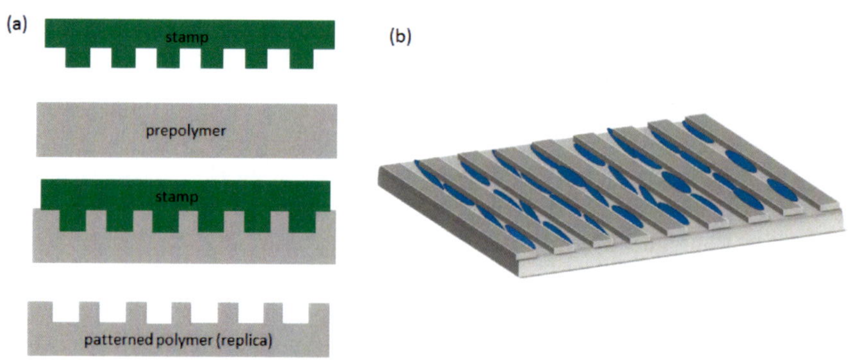

Figure 5. Patterned surface as LC alignment layer made by soft lithography. (a) Schematic of the fabrication steps for making patterned surface with an elastomeric stamp via replication process. (b) LC alignment on patterned surface with well-defined surface features of comparable dimensions to those of the mesogens.

For instance, a hybrid polymer consisting of two distinct moieties with largely different thermo-mechanical properties and surface activity was used in NIL. This material shows excellent capability both as a nanoimprinting material, requiring softness at moderate temperatures, and as a LC alignment layer, requiring sufficient rigidity. LC devices using the alignment layer show fairly stable electro-optic characteristics even after thermal aging, due to its high thermal stability. The soft component of the hybrid polymer helps it provide the multiple imprinting capabilities at higher temperatures virtually without any mold damages [170].

In another study, NIL was utilized to transfer a nano-pattern onto a homeotropic polyimide (PI) surface with the aid of nano-patterned silicon mold. The resulting nano-patterned PI layer acted as the homeotropic alignment layer, which uniformly aligned LC molecules in a vertically aligned LCD [171].

Soft lithography was employed for fabrication of patterned polymer films with variable feature dimensions, particularly one-dimensional polymer line gratings. The LC alignment on such patterned layers was found to be dependent on the amplitude and periodicity of topographic features, but also to be influenced by the mold removal step (peel-off direction). The surface modification occurs during peel-off and it was found to be dependent on the structure of the embossed polymer, and the chemical composition of the mold. These findings can be useful for finding alternate ways to affect LC alignment in a potential roll-to-roll manufacturing. The authors suggested that the physical topography was sufficient to align the LC director, but the chemical interactions between the mold and the alignment layer during the fabrication step, coupled with the peel-off step can affect the orientation of the LC director [175,176]. Also, alignment layer patterns exhibiting four-fold, six-fold and five-fold symmetries, which yield multi-stable LC alignment, were proposed. It is known that LC devices in which the LC molecules can adopt multiple stable orientations could significantly reduce the power consumption required for high-information-content displays [175].

In another study, a different soft lithographic technique, particualrly contact printing method with poly(dimethylsiloxane) (PDMS) stamp, was used for fabrication of a compartmentalized multidomain alignment layer. The ultrathin alignment layer consists of a self-assembled oligomer layer of poly(dimethylsiloxane) (PDMS) formed by utilizing the oligomers that diffuse out from a PDMS elastomer stamp during a contact printing process [185].

The effects of three different shapes of imprinted alignment layer, *viz.* U-shaped, V-like- and square-like grooves on the molecular orientation of LC, were studied. It was found that the shape of grooved alignment layer has a significant influence on the molecular orientation of LC. Among the studied patterns, U-shaped groove pattern resulted in a highest performance of LC orientation, pretilt angle and anchoring energy, comparable to that of conventionally-rubbed layer. Most likely, the smoothness of the groove for the continuous flow of LC and the ridge between the grooves significantly influences the LC alignment behavior. It is found that the shape of grooved alignment layer has a significant influence on the molecular orientation of LC [189].

Grating LC alignment has been of interest, because of its potential application in bistable nematic LCDs. Gratings were etched into PI alignment film using excimer laser exposure through a phase mask, which yielded homogeneous LC alignment with molecules aligned along the grooves of the grating. No pretilt was observed due to the grating symmetry; however, exposure of the PI film with elliptically polarized light at oblique incidence resulted in pretilted LC alignment [138].

After illumination by single-beam polarized pulse laser with doses below the ablation threshold energy, periodic microstructure was prepared on azobenzene-containing poly(urethane-imide) (PUI) film surface. The azobenzene group tended to be oriented perpendicular to the surface periodic microstructure. The illuminated PUI surface was then used as alignment layer for mesogens, which aligned closely along the surface microgrooves, but with weak anchoring energy, which indicated that LC alignment was controlled by both, the surface microgrooves and the anisotropic orientation of the azobenzene groups [190].

The influence of direct laser writing in PIs on the alignment of LC was investigated to generate complex patterns of locally uniaxial alignment zones. Laser writing was performed below the ablation threshold, which resulted in local melting of the PI. A surface relief structure was obtained as a result of the fast heating and cooling cycle during laser writing [193].

Local rubbing of the surface of a polymer using the stylus of an atomic force microscope (AFM), also termed *nano-rubbing* was suggested to avoid the scratches, as well as dust and debris generation associated with the conventional cloth rubbing of polymer films. Even with a small load force on the order of 1 nN, which does not generate any noticeable surface grooves (scratches), the surface showed the capability to uniformly align adjacent LCs. The unidirectional rubbing at high loads generated scratches on the PI surface, but when annealed at temperatures well below the glass transition or soaked into an organic solvent, these scratches relaxed and completely disappeared while maintaining the capability of aligning LCs [180].

In another study of surface patterning with AFM tip, it was shown that polymer chain alignment or grooves can be responsible for the LC alignment depending on the thickness of the polymer film. In thick polymer films (> 100 nm) the polymer chains will align in the direction of the scan due to the large lateral force exerted by the tip. In thin polymer films (5–20 nm), the polymer chains are strongly fixed to the surface and will not realign by the tip scan; instead grooves will be formed. The azimuthal surface anchoring the energy of thick polymer films does not depend on the scan line separation and scan force, is of the same order of magnitude as for conventionally rubbed polyimide surfaces, but increases by the number of scans [183].

A microscopic pattern exhibiting six-fold symmetry is inscribed on a PI surface using the stylus of an AFM. The hexagonal symmetry of the microscopic orientational domains in turn gives rise to three stable macroscopic LC orientations (tristable surface-aligned nematic LC), which are mutually switchable by an in-plane electric field. The resulting switching mode is surface-driven, and compatible for flexible displays [184].

Conductive AFM (CAFM) nanolithography was utilized to modify a silicon surface. Generated SiOx relief by CAFM provided good LC alignment control on micron and submicron scale and is a potential towards manufacturing high-resolution images with a smaller pixel size [179].

Despite the advantages over the conventional rubbing method, lithographic platforms, direct laser writing or AFM tip surface patterning are in the infancy stage and/or require further improvements for practical applications on large area substrates. However, in the past decade, there have been demonstration of e.g., continuous roller imprinting process, which presents a low cost, high throughput solution without the size limitation associated with batch imprinting.

2.5. Ion / Plasma Beam Liquid Crystal Alignment

The ion beam (IB) alignment technology utilizes ion particle or plasma beam, usually accelerated at high speed and scanning over organic or inorganic surfaces (Figure 6). The beam extreme accuracy allows fabrication of LCDs with high image uniformity and high contrast. In the past decade or so, a lot of research is carried out in this field as a promising alternative to the rubbing process, [194-224] which started with the pioneering work of the IBM researchers, who used low-energy ion beams impinging at a glancing angle on amorphous inorganic films, such as diamond-like carbon (DLC). They produced high quality laptop and desktop displays at relatively lower cost than the rubbing method [194]. The mechanism was explained by initial random atomic arrangement in the inorganic films and an "order" was induced by exposure to the IB, when "unfavourably" oriented rings of atoms were selectively destroyed. The planes of the remaining not destroyed rings of atoms are predominantly parallel to the direction of the IB. This induced surface "order" was responsible for LC alignment. NEXAFS spectroscopy was additionaly used to link the orientational bond order on a surface of IB-irradiated DLC films with the direction of LC alignment [195].

Figure 6. Surface anisotropic modification by directional ion/plasma beam bombardment.

Modification of DLC surface by Ar+ ion beam bombardment resulted in efficient LC alignment layer, comparable to that achieved on rubbed surface By changing IB conditions, such as the energy of the incident ions at incident angle of 45°, the LC pretilt angle was controlled. Comparable aligning performance to that achieved on rubbed surface was observed on such Ar-ion treated DLC surfaces [201].

Amorphous DLC thin films were also treated with plasma in various atmospheres and an increase in the fraction of sp2-bonding was observed after plasma treatments of the films. In particular, H_2 plasma-treated film had the largest fraction of sp2-bonding at the film surface and showed much improved alignment capabilities. These results suggest that p-bondings of the sp2-structure at the surface rather than the bulk play an important role in LC alignment [220].

Hydrogenated amorphous carbon (a-C:H) layers were submitted to IB irradiation, and even to atmospheric pressure plasma jet (APPJ), for their potential use as alignment LC layers. The novel APPJ technique offers a simple process with the potential in continuous in-line fabrication of LC alignment layers. Uniform alignment with a small pretilt angle was observed on surfaces treated with APPJ, which was attributed to the directional scanning treatment and formation of C-O and C=O bonds on the treated surfaces [223,224].

Also, a nitrogen plasma beam scanning treatment of a-C:H films yielded promising LC alignment layers. Despite the formation of grooves, which were considered not to be responsible for the LC alignment, the bond-breaking of aromatic rings which occurs on the surface of *a*-C:H films was associated with LC orientation. More specifically, the bond-breaking process of aromatic rings created carbon dangling bonds "convenient" for the formation of C–O bonds, which are "directional" and responsible for LC alignment [200]. The same research group also developed a two-step treatment by Ar and hydrogen plasma beam of a-C:H films, where the pretilt angle can be tuned by the two-step scanning treatment. The carbon dangling bonds were generated through a bond-breaking process during the first Ar plasma beam scanning step. The second hydrogen plasma beam scanning step was found to passivate some of the carbon dangling bonds generated in the first step and the pretilt angle decreased with the time of the second hydrogen plasma beam step.The decrease of pretilt angle was associated to the formation of C–H bonds that limit the formation of C–O bonds in air [205].

Besides carboneous surfaces, IB irradiation has been used tried on various inorganic layers, such as TiO_2, Al_2O_3, ZrO_2, ZnO, SiOx, HfO_2 and so on [208,213-215].

Amorphous silicon oxide (a-SiOx) and hydrogenated amorphous silicon oxide (a-SiOx:H) were irradiated using an Ar+ IB and tested for LC alignment. On the basis of experimental results, a-SiOx films deposited by sputtering do not align LC, but a-SiOx:H films treated with varying IB energies, IB incident angles, IB doses, and IB irradiation times have excellent alignment properties and electro-optical properties, identical to those rubbed PI. Additionally, hydrogen played an important role in LC alignment because of the difference in alignment properties between a-SiOx films and a-SiOx:H films. The proposed mechanism for LC orientation by IB-treated inorganic alignment layers was connected to the molecular interactions, such as van der Waals interaction, more than by physical topographical effects [226].

A low-energy IB exposure of SiOx films showed vertical or homogeneous LC alignment, as determined by electric force microscopy data. The authors associated the alignment to the Coulomb interactions between LC molecules and inorganic film surfaces [197].

LC alignment method that can align LC molecules homogeneously on IB-exposed indium–tin-oxide (IB-ITO) surface has been studied. A groove pattern wass formed parallel to the IB exposure direction. This method opens the possibility of using the ITO layer as LC alignment layer and as conductive layer (electrode) in LCD devices [229].

Yaroshchuk et al. described a new plasma technique yielding uniform planar or tilted orientation of a nematic LC on different organic and inorganic substrates. The method is based on the oblique irradiation of an aligning substrate with a “sheet” of accelerated plasma. Two modes of uniform alignment were observed, parallel and perpendicular to the incident plane of plasma beam, which were easily controlled with the irradiation dose, while the pretilt angle was controlled by changing the parameters of irradiation, viz. incidence angle, current density, particle energy, etc. [216-219].

Among the polymer surfaces subjected to IB bombardment, PI-surfaces are the most explored [198,199,203,212,222]. LC aligned homogeneously on IB-exposed PI films and the LC electro-optic switching behavior was similar to that of a cell containing rubbed layers. However, surface morphology and bonding molecules of IB-treated PI surfaces show properties very different from mechanically-rubbed ones. Particualrly, IB-treated PI surfaces results in the formation of hydrogenated amorphous carbon-like structure with short main-chains. In another study, Ar atomic beam treatment of PI films allows LC alignment parallel to the beam direction. The alignment mechanism was explained by the selective destruction of the double bonds in PI [222]. In yet another study, various pretilt angles were created on IB-irradiated PI surfaces and no changes in the surface morphology were observed. The surface chemical structure analysis showed that the alignment ability of LCs was due to the preferential reorientation of the carbon network due to the breaking of C=O double bonds in the imide ring parallel to the incident IB direction. Superior LC electro-optical characteristics on IB-treated PI layers compared to those measured on rubbed PI, were observed. Assumption was made that the LC direction depends strongly on the C–O bonds created from C=O bonds on the PI surface broken by IB irradiation.

Two-directional plasma beam processing was also used to investigate the orientation of LC molecules on PI surfaces. A phenomenon of rewriting LC alignment was noticed in two-directional plasma beam processing: the effect of the first plasma beam scanning was erased by the second one [209].

PI surface has been also treated with APPJ, which showed increase in C-O bonds. Anisotropic polyimidization method was reported for alignment surface coated with polyamic acid and IB-treated before polyimidization. LC molecules were homogeneously aligned along the direction of the polyimidization. The LC cells fabricated by this method have lower pretilt angle and stronger anchoring energy than the cells fabricated by standard IB method and rubbing method [199].

PDMS layer has been shown as promising pretilt-controllable alignment layer, which can provide planar or vertical alignment by varying the ion beam (IB) energy [196]. Also, LC pretilts angles were controlled over the range 0 - 90° using IB-treated substituted PS films as alignment layers [207].

A combination of atmospheric plasma treatment and rubbing process has been proposed for devices where the LC pretilt angle and anchoring energy need to be patterned in small areas or continuously tuned [201,204,209].

2.6. Novel Liquid Crystal Alignment Concepts

Various novel LC alignment approaches and trends will be briefly reviewed in this section. Not all excellent reports on novel LC alignment technologies will be mentioned in this chapter, but rather a variety in approaches was choosen. Alignment-layer-free approach, "self-alignment" methodology, "free surface" alignment control, alignment layers using carbon nanotubes, patterned SAMs, and nano-engineered surfaces are briefly described below [225-250].

Alignment-layer-free approach based on making patterns in a substrates, used in LCD device construction, has been described by Chao et al. Thermal nanoimprint lithography (thermal-NIL) has been utilized to transfer the microgroove pattern onto the plastic substrate. Without coating an alignment layer, the microgrooves on the flexible substrates can align LCs directly. The flexible LC cell maintained the electro-optical properties while bending. Also, the cracks observed in cells where additional alignment layer were coated during the bending process, can be avoided. Moreover, this method can be applied to the roll-to-roll process to increase the production efficiency and lower the cost [225,226].

Yamamoto et al. introduces the *"self-alignment" concept* to anchor and align nematic LC molecules homogeneously. The concept is based on the fact that if the glass-transition temperature of a polymer layer is low enough with respect to the nematic-isotropic transition temperature of the LC material, then the director of LC layer in contact with such polymer, can freely rotate on the "liquid "polymer layer coated over the substrate at a sufficiently high temperature. Therefore, unidirectional LC director alignment can be easily governed by the application of a very weak external field, then cooled down the LC cell, while keeping the induced LC orientation. As a result, the LC molecules are *anchored* and simultaneously *aligned* in melted polymer layer, which is then quenched. It is suggested that the "anchoring" force to fix LC molecules on the polymer layers originates from the local molecular interaction between LC molecules and the polymer. The anchoring energy of the "self-alignment" was almost comparable to that measured on conventional rubbed surface [227].

A new approach to control the orientation of LCs from the free (air) surface side, and not from the substrate side, has been proposed by Seki et al. A *"skin layer" at the free surface* of LC layer was prepared using a non-photoreactive LCP film by e.g., surface segregation or inkjet printing an azobenzene-containing LC-block copolymer. With a substrate on the other side, the skin layer generated both, planar-planar and homoeotropic-planar mode patterns. This approach offers the potential for LC-based devices with no modification of the substrate surface, any substrate to be used, while the LC layer orientation occurs via the free surface [228].

Surface nano-patterned indium–tin oxide (ITO) layer with high resolution and high aspect ratio, have been fabricated by a method based on the secondary sputtering phenomenon. The ITO patterned surfaces developed in this way not only provided high and stable anchoring energy comparable to that of rubbed PI, but also maintained its low resistivity to be used as an electrode. As a result, the patterned ITO layer has a dual role serving as a bifunctional conductive and alignment layer in LC devices [172]. Also, LC alignment method that can align LC molecules homogeneously on IB-exposed ITO surface without coating any additional alignment layer was mentioned in the previous section [229].

A highly aligned carbon nanotube (CNT) layer oriented crosslinked LCP material through a simple melting process. The resulting composite film can be rapidly bent and unbent by alternate irradiation with UV and visible light. The film also exhibits excellent mechanical and electrical properties due to the incorporation of aligned CNTs [230]. A method for aligning mesogens using magnetic-field oriented *single-wall carbon nanotubes* (SWNTs) modified with magnetic particles was described by Jung et al. A high degree of homeotropic/planar LC alignment was achieved by SWNTs exposed to a very low magnetic field due to the strong π–π interactions between the biphenyl group in the mesogens and SWNT's wall [231]. SWNTs as LC alignment materials have been also explored by Samulski et al. Particulalry, SWNT bundles were deposited on glass substrates with controlled nanotube orientation by two methods, viz. self assembly and dip-coating, generating dense and sparse SWNT films, respectively. These films were capable to initiate planar alignment of nematic LC [232,233].

Komitov suggested that nano-scale changes in the alignment layers can affect the anchoring strength and orientation of the overlaying LC layer. Based of nano-engineering of surface alignment properties, two novel alignment concepts have been introduced: *electrically commanded surfaces (ECS)* and *high-performance alignment layers (HiPAL).* ECS materials are actually active alignment materials designed to switch LC orientation, while HiPAL materials enable control of the anchoring strength and pretilt angles in a broad range, from 0° - 90° [236,249]. In ECS design, the LC layer is aligned by means of thin ferroelectric liquid-crystal-polymer film. An electric field applied normal to the film, switches the molecules of the ferroelectric film, which via elastic forces, further transfers to the LC layer [249].

Another nano-structured alignment layers with controllable polar and azimuthal anchoring energies have been proposed by Kwok et al. The alignment layer is based on a nanostructured surface formed by mixing a conventional PI and a nonaligning polymer. By using this variable anchoring alignment layer, bistable LCDs are possible [250].

Self-assembled monolayers (SAMs) and Langmuir-Blodgett (LB) films with their uniformity, controllable thickness and reproducible properties, have been extensively explored as LC alignment layers. For instance, Nolte et al. developed a simple method for fabrication of a stable, *tunable self-assembled alignment layer* based on pyridine-functionalized oligosiloxane, which spontaneously forms an anisotropic, grooved surface on ITO layer.

The pyridine acts as seeds for the epitaxial growth of stacks of highly ordered zinc phthalocyanines. By varying the height of the stacks, direct control over the properties of LC domains was achieved [234,235]. LC cells fabricated with patterned SAM layers on ITO showed superior electro-optical performance compared to conventional TN-cells: lower operating voltages, faster switching speeds and no flow-induced dynamic cross-talk between neighboring switching LC domains. The pattern-symmetry-induced multistability that is intrinsic to this method opens new routes for design and fabrication of numerous LC devices, e.g., multistable LC devices (by using in-plane electrodes) and switchable viewing-angle displays [245].

Many other technologies, like friction transfer, polymer film stretching, surface wrinkle method or ink-jet printing for making topographical features, and many monolayer and multilayer designs (SAM, LB, layer-by-layer) have been prposed for LC alignment, but were not covered in this chapter.

CONCLUSION

An overview of alignment technologies for nematic LCs proposed and developed in the past two decades is given. Emphasis was put on the origins of LC alignment on surfaces of different alignment layers, such as rubbed, photo-irradiated, obliquely-deposited, IB-bombarded and patterned surfaces. Numerous studies, experimental and theoretical, are in favor of surface-mesogen interactions as main contributor to the surface preferential LC orientation, whereas the surface topology (if present) has a minor contribution; only when the molecular interactions were weak, the contribution from the topography was noticeable. In such cases, the relative contributions of topology effects and molecular interactions on LC alignment have not yet been quantified in detail.

While numerous attempts to find suitable alternative to the rubbing method have been made, the LCD industry is still mostly relying on the rubbing method, a method proposed a century ago. Until recently, after almost 30 years extensive research in the field LC photoalignment, one of the largest LCD manufacturers adopted the photo-induced LC alignment for making high resolution bright LCD panels. It remains to see if the rubbing method will be completely replaced in the years to come, especially with the appearance of novel LC alignment methodologies, and some of them, even do not need a separate alignment layer.

REFERENCES

[1] Takatoh, K.; Sakamoto, M.; Hasegawa, R.; Koden, M.; Itoh, N.; Hasegawa, M. *Alignment Technology and Applications of Liquid Crystal Devices*; Taylor and Francis Group, 2005.

[2] Crawford G. *Flexible Flat Panel Displays;*Wiley-Series in Display Technologiy, 2005.

[3] Naoyuki, K. The Liquid Crystal Display Story: 50 Years of Liquid Crystal R&D that lead The Way to the Future; Springer, 2014.

[4] Hahm, S. G.; Ko, Y.-G.; Rho, Y.; Ahn, B.; Ree, B. *Current Opinion in Chemical Engineering* 2013, *2*, 71–78.

[5] Yang, D.-K.; Wu, S.-T. *Fundamentals of Liquid Crystal Devices*; Wiley-Series in Display Technologiy, 2nd Ed., 2014.

[6] Chen, J.; Cranton, W.; Fihn, M. *Handbook of Visual Display Technology;* Springer, 2012, 1387-1402.

[7] Kelly, S. M. *Flat Panel Displays: Advanced Organic Materials;* Royal Society of Chemistry, Cambridge, UK, 2000.

[8] Atkuri H. M.; Leong, E. S. P.; Hwang, J.; Palermo, G.; Si, G.; Wong , J.-M.; Chien, L.-C.; Ma, J.; Zhou, K.; Liu, Y. J.; De Sio, L. *Displays* 2015, *36*, 21–29.

[9] Rasing, T.; Musevic, I. Surfaces and Interfaces of Liquid Crystals; Springer, 2004.

[10] Chigrinov, V. G.; Kozenkov, V. M.; Kwok, H.-S. *Photoalignment of Liquid Crystalline Materials: Physics and Applications;* Wiley-Series in Display Technologiy, 2008.

[11] Mauguin, C. V. *Bull. Soc. Fr. Mineral.* 1911, *34*, 71–117.

[12] Chatelain, P. *Bull. Soc. Fr. Mineral.* 1944, *88*, 105–130.

[13] Hoogboom, J.; Rasing, T.; Rowan, A. E.; Nolte, R. J. M. *J. Mater. Chem.* 2006, *16*, 1305–1314.

[14] Mosley, A.; Nicholas, B. M.; Gass, P. A. *Displays* 1987, *8*(1), 17-21.

[15] Yang, F.; Zoriniants, G.; Ruan, L.; Sambles, J. R. *Liquid Crystals* 2007, 34(12), 1433–1441.

[16] Chang, C. J.; Chou, R.-L.; Lin, Y.-C.; Liang, B.-J.; Chen, J.-J. *Thin Solid Films* 2011, *519*, 5013–5016.

[17] Hahm, S. G.; Lee, S. W.; Suh, J.; Chae, B.; Kim, S. B.; Lee, S. J.; Lee, K. H.; Jung, J. C.; Ree, M. *High Performance Polymers* 2006, *18*,549–577.

[18] Lee, T. J.; Hahma, S. G.; Lee, S. W.; Chae, B.; Lee, S. J.; Kima, G.; Kima, S. B.; Jung , J. C.; Ree, M. *Materials Science and Engineering B 132* 2006, 64–69.

[19] Ishihara, S.; Wakemoto, H.; Nakazima, K.; Matsuo, Y. *Liquid Crystals* 1989, *4*(6), 669-675.

[20] Lee, S. W.; Chae, B.; Kim, H. C.; Lee, B.; Choi, W.; Kim, S. B.; Chang, T.; Ree, M. Langmuir **2003**, *19* (21), 8735–8743.

[21] Cui, Y.; Zola, R. S.; Yang, Y.-C.; Yang, D.-Ke *J. Appl. Phys.* 2012, *111*, 063520.

[22] Kumar, T. A.; Takezoe, H.; Dhara, S. Japanese Journal of Applied Physics 2011, *50*, 040203.

[23] Garbovskiy, Y.; Reisman, L.; Celinski, Z.; Camley, R. E.; Glushchenko, A. *App. Phys. Lett.* 2011, *98*, 073301

[24] Fang, Yu-Q.; Wang, J.; Zhang, Q.; Zeng, Y.; Wang, Y.-H. *European Polymer Journal* 2010, *46*, 1163.

[25] Wang, J.; Wang, Y. H. Chinese Chemical Letters 2008, *19*, 359–362.
[26] Liu, Z.; Yu, F.; Zhang, Q.; Zeng, Y.; Wang, Y. *European Polymer Journal* 2008, *44*, 2718–2727.
[27] Seo, D.-S. Journal of the Korean Physical Society 2000, 36(1), 29-33.
[28] Chang, C.-J.; Tseng, I-H..; Lin, Y.-C.; Wu, B. S.; Liang, B.-J.; Tsai, M.-H. *Thin Solid Films* 2013, *544*, 74–78.
[29] Liu, J.-G.; Zhao, X.-J.; LI, H.-S.; Fan, L.;, Yang, S.-Y. *High Performance Polymers* 2006, *18*, 851–865.
[30] B Jerome, B. *Rep. Prog. Phys.* 1991, *54*(3), 391.
[31] Stohr, J.; Samant, M. G. Journal of Electron Spectroscopy and Related Phenomena 1999, *98–99,* 189–207.
[32] Stohr, J.; Samant, M. G.; Cossy-Favre, A.; J. Dıaz, J.; Momoi, Y.; Odahara, S.; Nagata, T. *Macromolecules* 1998, *31,* 1942-1946.
[33] Kumar, S.; Jae-Hoon Kim, J.-H.; Shi, Y. *Phys. Rev. Lett.* 2005, *94*, 077803.
[34] Matveenko, V. N; Kirsanov, E. A. *Russ. Chem. Rev.* 1986, *55,* 743.
[35] Snively, C. M.; Koenig, J. L. *Journal of Molecular Structure* 2000, *521*, 121–126.
[36] Jiao, M.; Ge, Z.; Song, Q.; Wu, S.-T. *Appl. Phys. Lett.* 2008, *92*, 061102.
[37] Huang, C.-Y.; Huang, Y.-S.; Jing-Rui Tian, J.-R.; Yeh, S.-J. *Optics Communications* 2006, *265*, 241–245.
[38] Berreman, D. W. Phys. Rev. Lett. 1972, *28*, 1683.
[39] Fukuda, J.; Yoneya, M.; Yokoyama, H. *Phys Rev Lett.* 2007, *98*(18), 187803.
[40] Fukuda, J.; Gwag, J. S.; Yoneya, M.; Yokoyama, H. *Phys. Rev. E Stat. Nonlin. Soft Matter. Phys.* 2008, *77*, 011702.
[41] Geary, J. M.; Goodby, J. W.; Kmetz, A. R; Patel, J. S. *J. App. Phys*. 1987, 62(10), 4100-4108.
[42] Tsuda , Y.; Jae Min OH, J. M.; Kuwahara, R. *Int. J. Mol. Sci.* 2009, *10*, 5031-5053.
[43] Kim, D.-H.; Park, H.-G.; Kim, Y.-H.; Kim, B.-Y.; Ok, C.-Ho; Seo, D.-S. Transactions on Electrical and Electronic Materials 2009, 10(4).
[44] Lee, Y.-J.; Gwag, J. S.; Kim, Y.-K.; Jo, S. I.; Kang, S.-G.;Young Ran Park, Y. R.; Kim, J.-H. *App. Phys. Lett.* 2009, *94*, 041113.
[45] Kang, H.; Park, J. S.; Sohn , W.-H.; Kang, D.; Rosenblatt, C.; Lee, J.-C. *Polymer* 2009, *50*, 5220–5227.
[46] Jeng, S.-C.; Kuo, C.-W.; Wang, H.-L.; Liao, C.-C. *Appl. Phys. Lett.* 2007, *91,* 061112.
[47] Hwang, S.-J.; Jeng, S.-C.; Yang, C.-Y.; Kuo, C.-W.; Liao, C.-C. *J. Phys. D: Appl. Phys.* 2009, *42,* 025102.
[48] Hwang, S.-J.; Jeng, S.-C.; Hseih, I.-M. *Opt. Express* 2010, *18*(16),16507–16512.
[49] Sung, S.-J.; Jung, E. A.; Kim, D.-H.; Kang, J.-K.; Yang, K.-J.; Do, Y. S.; Choi, B.-D. *Displays* 2010, *31*, 93–98.
[50] Son, P. K.; Park, J. H.; Kim, J. C.; Yoon, T.-H. *Thin Solid Films* 2007, *515,* 3102–3106.
[51] Chen, C.; Bos, P. J.; Anderson, J. E. *Liquid Crystals* 2008, *35*(4), 465–481.
[52] Huang, Y.; Bos, P. J.; Bhowmik, A. *J. App. Phys.* 2010, *108*, 064502.
[53] Sung, S. J.; Yang, K. J.; Kim, D. H.; Do, Y. S.; Kang, J. K.; Choi, B. D.; *J. Nanosci. Nanotechnol.* 2009; *9*(12), 6938-42.
[54] Kim, K. C.; Ahn, H. J.; Kim, J. B.; Hwang, B. H.; Baik, H. K. *Langmuir* 2005, *21*(24), 11079-84.

[55] Hwang, B. H.; Ahn, H. J.; Rho, S. J.; Chae, S. S.; Baik, H. K. *Langmuir* 2009, *25*(14), 8306-12.
[56] Liou, W.-R.; Chen, C.-Y.; Ho, J.-J.; Hsu, C.-K.; Chung-Cheng Chang, C.-C.; Hsiao, R. Y.; Chang, S.-H. *Displays* 2006, *27*, 69–72.
[57] Celinski, Z.; Reisman, L.; Harward, I.; Glushchenko, A. Proc. of SPIE Vol. 7329 73290G-1, 2009.
[58] Khakhlou, A.; Murauski, A.; Yaroshchuk, O.; Telesh, E.; Kravchuk, R. *J. Soc. Inf. Display* 2006, *14* (3), 257-263.
[59] Chen, M.-Z.; Chen, W.-S.; Jeng, S.-C.; Yang, S.-H.; Chung, Y.-F. *Optics Express* 2013, 21(24), 29277-29282.
[60] Lee, J.-J.; Han, J.-J.; Park, H.-G.; Kim, D.-H.; Byun, S.-U.; Seo, D.-S. *Optical Materials* 2013, *27*, 2658–2663.
[61] Hwang, S.-W.; Seo, J.-H.; Yoon, T.-H.; Kim, J. C. *Thin Solid Films* 2010, *519*, 885–889.
[62] Kim, B.-Y.; Kim, Y.-H.; Park, H.-G.; Seo, D.-S. *Liquid Crystals* 2010, *37*(11), 1381–1384.
[63] Hong, C.; Tang, T.-T.; Hung, C.-Y.; Pan, R.-P.; Fang, W. *Nanotechnology* 2010, *21*, 285201.
[64] Ichimura, K.; Suzuki, Y.; Seki, T.; Hosoki, A.; Aoki, K. *Langmuir* 1988, ***4***, 1214–1216.
[65] Gibbons, W. M.; Shannon, P. J.; Sun, S.-T.; Swetlin, B. J. *Nature* 1991, ***351,*** 49–50.
[66] Shannon, P. J.; W. M. Gibbons , W. M.; Sun, S. T. *Nature* 1994, *368*, 532-533.
[67] Gibbons, W. M.; Shannon, P. J.; Sun, S.-T. *Mol. Cryst. Liq. Cryst.* 1994, *251*(1), 191-208.
[68] Schadt, M.; Schmitt, K.; Kozinkov, V.; Chigrinov, V. *Jpn J. Appl. Phys.* 1992, ***31***, 2155–2164.
[69] Schadt, M., Mol. Cryst. Liq. Cryst. 2001; 364(1), 151-169
[70] Schadt, M.; Seiberle, H.; Schuster, A. *Nature* 1996, 381(6579), pp. 212-215.
[71] Seiberle, H.; Muller, O.; Marck, G.; Schadt, M. *Journal of the SID* 2002, 10/1, 31-35
[72] Newsome, C. J.; O'Neill, M. *J. Appl. Phys.* 2002, 92(4), 1752.
[73] Lu, J.; Deshpande, S. V.; Gulari, E.; Kanicki, J.; Warren, W. L. *J. Appl. Phys.* 1996, *80* (9), 5028-5034.
[74] Ichimura, K. *Chem. Rev.* 2000, *100*, 1847–1873.
[75] Seki, T.; Nagano, S.; Hara, M. *Polymer* 2013, *54*, 6053-6072.
[76] O'Neill, M.; Kelly, S. M. *J. Phys. D.* 2000, *33*, R67.
[77] Yaroshchuk, O.; Reznikov, Y. *J. Mater. Chem.* 2012, *22,* 286-300
[78] Ichimura, K. *J. Synth. Org. Chem.* 1991, *49*(5), P 383-391.
[79] Fukuda, K.; Seki, T.; Ichimura, K. *Macromolecules* 2002, ***35,*** 1951–1957.
[80] Kidowaki, M.; Fujiwara, T.; Ichimura, K. *Chem. Lett.* 1999, ***28,*** 643–644.
[81] Yua,Y.; Ikeda, T. Journal of Photochemistry and Photobiology C: Photochemistry Reviews 2004, 5, 247–265.
[82] Bobrovsky, A.; Ryabchun, A.; Shibaev, V. Journal of Photochemistry and Photobiology A: Chemistry 2011, 218, 137–142.
[83] Furumi, S.; Ichimura, K. *Thin Solid Films* 2006, *499,* 135 – 142.
[84] Yaroshchuk, O.; Ho, J.; Chigrinov, V.; Kwok, H.-S. *Jpn. J. Appl. Phys.* 2007, *46*(5A), 2995–2998.
[85] Seki, T. *Polymer Journal* 2014, *46*, 751–768.

[86] Nalwa, H. S. Handbook of Advanced Electronic and Photonic Materials and Devices, Vol. 7: Liquid Crystals, Display and Laser Materials; Academic Press, 2000.
[87] .Kwok, H.-S.; Chigrinov, V. G.; Takada, H.; Takatsu, H. *IEEE/OSA Journal of Display Technology* 2005,1(1), 41-50.
[88] Chigrinov, V.; Kwok, H. S.; Takada, H.; Takatsu, H. *Liquid Crystals Today* 2005,*14*(4), 1–15.
[89] Chigrinov, V. C. *Photonics Letters of Poland* 2010, *2* (3), 104-106.
[90] Yaroshchuk, O.; Ho, J.; Chigrinov, V.; Kwok, H.-S. *Jpn J. Appl. Phys.* 2007, 46(5A), 2995–2998.
[91] Chrzanowski, M. M.; Zieliński, J.; Olifierczuk, M.; Kędzierski, J.; Nowinowski-Kruszelnicki, E. *J. Achievements in Materials and Manufacturing Engineering* 2011, *48*(1), 7-13.
[92] Corvazier, L.; Zhao, Y. *Macromolecules* 1999, *32,* 3195-3200.
[93] Bartkiewicz, S.; Matczyszyn, K.; Mysliwiec, J.; Yaroshchuk, O.; Kosa, T.; Palffy-Muharey, P. *Mol. Cryst. Liq. Cryst.* 2004, 412, 301-312.
[94] Statman, D.; Page, E.; Werner, V.; Lombardi, J. C. Phys. Rev. E Stat. Nonlin. Soft Matter. Phys. 2007, *75*, 021703.
[95] Fang. G.; Shi, Y.; Maclennan, J. E.; Clark, N. A.; Farrow, M. J.; Walba, D. M. *Langmuir* 2010, *26*(22), 17482-8.
[96] Usami, K.; Sakamoto, K. J. *App. Phys.* 2011, *110*, 043522.
[97] Wu , Y.; Demachi, Y.; Tsutsumi , O.; Kanazawa, A.; Shiono , T.; Ikeda, T. *Macromolecules* 1998, *31* (4), 1104–1108.
[98] Wu, Y.; Kanazawa, A.; Shiono, T.; Ikeda, T.; Zhang, Q. *Polymer* 1999, *40,* 4787–4793.
[99] Cho, K.; Sung, S.; Lee, S.; Chang, H.; Park J.-K. *Materials Science and Engineering: C* 2004, 24(1), 195-199.
[100] Zhang, L.; Peng, Z.; Yao, L.; Fei, C.; Lv, F.; Xuan, L. Applied Surface Science 2007, *253,* 3372–3377.
[101] Shen, B. J.; Xie, Z.; He, Y. N.; Lian, Y. Q. *Chinese Chemical Letters* 2008, *19*, 1131–1134.
[102] Lee, J.; Lee, S.; Jeong, Y.- C.; Cho, K. Y.; Park, J.-K. *Optics Express* 2009, *17*(26), 23565-23575.
[103] Kim, C.; Trajkovska, A.; Wallace, J. U.; Chen, S. H. *Macromolecules* 2006, *39,* 3817-3823.
[104] Trajkovska, A.; Kim, C.; Marshall, K. L.; Mourey, T. H.; Chen, S. H. *Macromolecules* 2006, *39,* 6983-6989.
[105] Kim, C.; Wallace, J. U.; Trajkovska, A.; Ou, J. J.; Chen, S. H. *Macromolecules* 2007, *40,* 8924-8929.
[106] Kim, C. Photoalignment of Liquid Crystals and Development of Novel Glassy Liquid Crystals, PhD Thesis, University of Rochester, Rochester, NY, 2008.
[107] Trajkovska, A. Chrioptical Properties and Photoalignment of Monodisperse Glassy Liquid Crystalline Oligofluorenes, PhD Thesis, University of Rochester, Rochester, NY, 2006.
[108] Trajkovska, A. Liquid Crystalline Oligofluorenes: Chrioptical Properties and Photoalignment, VDM Verlag Dr. Mueller, 2008.
[109] Kim, K.; Wallace, J. U.;Chen, S. H.; Merkel, P. B. *Macromolecules* 2008, *41*, 3075.

[110] Wei, S. K.-H.; Zeng, L.; Marshall, K. L.; Chen, S. H. *J. Polymer Sci. Part B: Polymer Phys.* 2011, *49*, 725.
[111] Kawatsuki, N.; Kawanishi, T.; Uchida, E. Macromolecules **2008**, *41*(13), 4642–4650.
[112] Kawatsuki, N.; Takatsuka, H.; Yamamoto, T.; Ono, H. *Jpn. J. Appl. Phys.* 1997, *36,* 6464.
[113] Kawatsuki, N.; Koezuka, Y. *Polymer* 2009, *50*(11), 2349–2356.
[114] Kawatsuki, N.; Takatani, K.; Yamamoto, T.; Ono, H. *Polymer Journal* 1998, *30,* 946-949.
[115] Kawatsuki, N.; Takatsuka, H.; Kawakami, Y.; Yamamoto, T.; *Polymers for Advanced Technologies, Special Issue: Papers Presented at FCFP-VIII* 1999, *10*(7), 429–433.
[116] Kawatsuki, N.; Inada, Y.; Goto, K.; Kondo, M.; Haruyama, Y.; Matsui, S. *Mol. Cryst. Liq. Cryst.* 2014, 601(1).
[117] Cao, J.;, Zhang, W.-M.; Yang, J.; Pu, J.-L. *Imaging Science and Photochemistry* 2013, *31*(6), 456-462.
[118] Furumi, S.; Ichimura, K. *Phys. Chem. Chem. Phys.* 2011,*13*, 4919-4927.
[119] Yu, F. F.; Cao, J.; Zhang, W. M.; Pu, J. L. *Advanced Materials Research* 2012, *531*, 137-140.
[120] Reznikov, Y.; Petschek, R. G.; Rosenblatt, C. *Appl. Phys. Lett.* 1996, *68* (16), 2201-2203.
[121] Andrienko, D.; Kurioz, Y.; Reznikov, Y.; Rosenblatt, C.; Petschek, R.; Lavrentovich, O.; Subacius, D. *J. Appl. Phys.* 1998, *83* (1), 50-55.
[122] Rich, D. C.; Sichel, E. K.; Cebet, P. MRS Proceedings 1995, 413.
[123] Kim, H.-T.; Lee, J.-W.; Sung, S.-J.; Park, J.-K. *Liquid Crystals* 2000, *27*(10), 1343-1356.
[124] Sung, S.-J.; Kim, H.-J.; Lee, J.-W.; Park, J.-K. *Synthetic Metals* 2001, *117*, 277-279.
[125] Guo, M.; Wang, X. *European Polymer Journal* 2009, *45*, 888–898.
[126] Usami, K.; Sakamoto, K.; Yokota, J.; Uehara, Y.; Ushioda, S. *Thin Solid Films* 2008, *516,* 2652–2655.
[127] Oo, T. N.; Iwata, T.; Kimura, M.; Akahane, T. *Science and Technology of Advanced Materials* 2005, *6*, 149–157.
[128] Sung, S.-J.; Kim, H.-K.; Lee, J.-W.; Park, J.-K. *Synthetic Metals* 2001, *117*(1–3), 277–279.
[129] Sung, S.-J.; Cho, K.-Y.; Park, J.-K. *Materials Science and Engineering: C* 2004, *24* (1–2), 181–184.
[130] Hahm, S. G. Liquid Crystal Aligning and Nonvolatile Memory Characteristics of Polyimides in Nanoscale Thin Films, PhD Thesis, Korea, 2008.
[131] Nishikawa, M.; Taheri, B.; West, J. L. *Appl. Phys. Lett.* 1998, *72*(19), 2403-2405.
[132] Chung, D.-H.; Takanishi, Y.; Ishikawa, K.; Takezoe, H.; Park, B.; Jung, Y.; Hwang, H.-K.; Lee, S.; Han, K.-J.; Jang, S.-H. *Jpn. J. Appl. Phys.* 2000, *39*, 185–187.
[133] Andrienko, D.; Kurioz, Y.; Nishikawa, M.; Reznikov, Y.; West, J. *Jpn. J. Appl. Phys.* 2000, *39*, 1217–1220.
[134] Yamaguchi, R.; Goto, Y.; Sato, S. *Jpn. J. Appl. Phys.* 2002, *41*, 889–891.
[135] Yamaguchi, R.; Sato, S. *Appl. Phys. Lett.* 2005, *86*, 031913.
[136] Chaudhari, P.; Lacey, J.; Lien, S. A.; Speidell, J. *Jpn. J. Appl. Phys.* 1998, *37*, 55–57.
[137] Pan, R.-P.; Chiu, H.-Y.; Lin, Y.-F.; Huang, J. Y. *Chinese Journal of Physics* 2003, *41*(2), 177-184.

[138] Newsome, C. J.; O'Neill, M.; Bryan-Brown, G. P. *Proc. SPIE 3618, Laser Applications in Microelectronic and Optoelectronic Manufacturing IV*, 1999.
[139] Li, X. T.; Natansohn, A.; Rochon, P. *Appl. Phys. Lett.* 1999, *74*, 3791.
[140] Miyachi, K.; Kobayashi, K.; Yamada, Y.; Mizushima, S. *SID Symposium Digest of Technical Papers* 2010, *41(1),* 579–582.
[141] Ibn-Elhaj, M.; Chappellet, S.; Lincker, F. *Solid State Phenomena* 2012, *181-182,* 3-13.
[142] Galabova, H. G.; Allender, D. W.; Chen, J. *Phys. Rev. E* 1997, *55*, 1627.
[143] Osterman, J.; Skarp, K. *SID* 2004, 35(1).
[144] Zhang, C.-H.; Yang, Z.-H.; Ding, M.-X. *Liq. Cryst.* 2003, *30*(1), 65-69.
[145] Andrienko, D.; Dyadyusha, A.; Kurioz, Y.; Reshetnyak, V. Y.; Reznikov, Y. A. *Proc. SPIE* 3488, *Nonlinear Optics of Liquid and Photorefractive Crystals* II, 175, 1998.
[146] Reznikov, Y.; Yaroshchuk, O.; Gerus, I. *Mol.Materials* 1998, *9,* 333-342.
[147] Yaroshchuk, O.; Pelzl, G.; Pirwitz, G.; Reznikov, Y.; Zaschke, H.; Kim, J. H.; Kwon, S. B. *Jpn. J. Appl. Phys.* Part 1 1997, *36* (9A), 5693-5695.
[148] Yaroshchuk, O.; Cada, L.G.; Sonpatki, M.; Chien, L.-C. *Appl.Phys.Lett.* 2001, *79*(1), 30-32.
[149] Vretik, L.; Syromiatnikov, V.; Zagniy, V.; Paskal, L.; Yaroshchuk, O.; Dolgov, L.; Kyrychenko, V.; Lee, C-D. *Mol.Cryst.Liq.Cryst.* 2007, *479*, 121-134.
[150] Hah, H.; Lee, J.; Sung, S.-J.; Cho, K. Y.; Park, J.-K. *Microelectronic Engineering* 2012, *96*, 67–70.
[151] Ogawa, K.; Ohtake, T.; Nomura, T. *Jpn. J. Appl. Phys.* 2000, *39*, 5904.
[152] Lee, J.-H.; Kang,D. *Proc. SPIE 8279, Emerging Liquid Crystal Technologies VII, 827919*, 2012.
[153] Zhu, J.; Wu, L.; Shi, K.; He, S.; Zhang, P.; Guo, X.; Du, Z.; Zhang, B. *Reactive & Functional Polymers* 2007, *67,* 1–9.
[154] Vretik, L.; Paskal, L.; Syromyatnikov, V.; Zagniy, V.; Savchuk, O.; Dolgov, L.; Yaroshchuk, O.; Lee, C.-D. *Mol.Cryst.Liq.Cryst.* 2007, *468*, 173–179.
[155] Alla, R. A.; Hegde, G.; Komitov, L. *Appl. Phys. Lett.* 2013, *102*, 233505.
[156] Alla, R. A. *On the Control of Nematic Liquid Crystal Alignment*, PhD thesis, University of Gothenburg, 2013.
[157] Gerus, I.; Glushchenko, A.; Kwon, S.-B.; Reshetnyak, V.; Reznikov, Y. *Liquid Crystals* 2001, *28*(11), 1709-1713.
[158] Thieghi, L.T.; Bonvent, J.J.; Oliveira, E.A.; Giacometti, J.A.; Balogh, D.T. *Applied Physics A* 2003, *77(7),* 911-914.
[159] Xu, C.; Shiono, T.; Ikeda, T.; Wang, Y.; Takeuchi, Y. *J. Mater. Chem.* 2003, *13*, 669-671.
[160] Ree, M.; Kim, S.; Lee, S. W. *Synthetic Metals* 2001, *117*, 273-275.
[161] Jeng, S-C.; Hwang, S.-J.; Chen, T.-A.; Liu, H.-S.; Chen, M.-Z. *Proc. SPIE* 8279, *Emerging Liquid Crystal Technologies* VII, 827912, 2012.
[162] Liu, J.; Liang, X.; Tang, H.; Xu, S.; Gao, H. *Thin Solid Films* 2001, 384(2), 212-214.
[163] Jung, K. H.; Hyun, S.-Y.; Song, D.-M.; Shin, D.-M. *Optical Materials* 2002, *21*, 663–666.
[164] Li, X.-D.; Zhong, Z.-X.; Jin, G.; Lee, S. H.; Lee, M.-H. *Macromolecular Research* 2006, *14*(3), 257-260.
[165] Faetti, S.; Sakamoto, K.; Usami, K. *Phys. Rev. E Stat. Nonlin. Soft Matter. Phys.* 2007, *75*(5 Pt 1), 051704.

[166] Mahilny, U. V.; Stankevich, A. I.; Muravsky, A. A.; Murauski, A. A. *J. Phys. D: Appl. Phys.* 2009, *42*, 075303.

[167] Kang, H.; Kang, D.; Lee, J.-C. *Polymer* 2009, *50*, 2104–2112.

[168] Fei, C.-H.; Peng, Z.-H.; Lv, F.-Z.; Zhang, L.-L.; Yao, L.-S.; Xuan, L. *Chinese Phys. Lett.* 2006, *23*, 3029.

[169] Vengatesan, M.R.; Lee, S.H.;Son, J.H.; Lim,J.K.; Song, J.K.; *J.Colloid and Interface Sci.* 2013, *407,* 310-317.

[170] Gwag, J. S.; Oh, M.; Yoneya, M.; Yokoyama, H.; Satou, H.; Itami, S. *J. Appl. Phys.* 2007, *102*, 063501.

[171] Park, H.-G.; Lee, J.-J.; Dong, K.-Y.; Oh, B.-Y.; Kim, Y.-H.; Jeong, H.-Y.; B.-K.; Seo, D.-S. *Soft Matter.* 2011, *7*, 5610-5614.

[172] Jeong, H. S.; Jeon, H.-J.; Kim, Y. H.; Oh, M. B.; Kumar, P.; Kang, S.-W.; Jung, H.-T. *NPG Asia Materials* 2012, *4* (e7), 1-7.

[173] Choi, Y.; Yokoyama, H.; Gwag, J. S. *Optics Express* 2013, 21(10), 12135-12144.

[174] Lin, R.; Rogers, J. A. *Nano Lett.* 2007, *7*(6), 1613-1621.

[175] Schenck, W. A. *Liquid Crystal Alignment on Embossed Polymer Films,* PhD thesis, University of North Carolina at Chapel Hill, 2010.

[176] Schenck, W.; Ko, D.-H.; Samulski, E. *J. Appl. Phys.* 2011, *109*, 064301.

[177] Hah, H.; Sung, S.-J.; Han, M.; Lee, S. S.; Park, J.-K. *Displays* 2008, *29*, 478–481.

[178] Crawford, G. P.; Eakin, J. N.; Radcliffe, M. D.; Callan-Jones, A.; Pelcovits, R. A. *J. Appl. Phys.* 2005, *98*, 123102.

[179] Lin, T.-C.; Chao, C.-Y. *Mol. Cryst. Liq. Cryst.* 2009, *510*, 134–140.

[180] Kim, J.-H.; Yoneya, M.; Yamamoto, J.; Yokoyama, H. *Nanotechnology* 2002, *13,* 133–137.

[181] Zheng, W. J.; Huang, M.H. *Thin Solid Films* 2012, *520*, 2841–2845.

[182] Scharf, T.; Shlayen, A.; Gernez, C.; Basturk, N.; Grupp, J. *Mol. Cryst. Liq. Cryst.* 2004, *411*, 135–145.

[183] Rastegara, A,; Skarabot, M.; Blij, B.; Rasing, T. *J. Appl. Phys.* 2001, 89(2), 960-964.

[184] Kim, J.-H.; Yoneya, M.; Yokoyama, H. *Nature* 2002, *420*, 159-162.

[185] Chae, S. S.; Min, H.; Lee, J. H.; Hwang, B.; Sung, W. M.; Jang, W. S.; Yoo, Y. B.; Oh, J.; Park, J. H.; Kang, D.; Kim, D.; Kim, Y. S.; Baik, H. K. *Adv. Mater.* 2013, 25(10), 1408–1414.

[186] Lee, B. W.; Clark, N. A. *Science* 2001, *291*(5513), 2576-80.

[187] Melik, H.; Emek, M.; Ylmaz, S. Cankaya University Journal of Science and Engineering 2011, 8(1), 99-109.

[188] Son, J.-H.; Zin, W.-C.; Takezoe, H.; Song, J.-K. *Adv. Mater.* 2012, *24*(45), 6105–6110.

[189] Hah, H.; Sung, S.-J.; Han, M.; Lee, S.; Park, J.-K. *Materials Science and Engineering* 2007, C27, 798–801.

[190] Lu, X.; Lu, Q.; Zhu, Z.; Yin, J.; Wang, Z. *Chemical Physics Letters* 2003, *377*, 433–438.

[191] Kagajyo, T.; Fujibayashi, K.; Shimamura, T.; Okada, H.; Onnagawa, H. *Jpn. J. Appl. Phys.* 2005, 44(1B), 578–581.

[192] Torii, S.; Makimura, T.; Okazaki, K.; Nakamura, D.; Takahashi, A.; Okada, T.; Niino, H.; Murakami, K. *Applied Physics Express* 2010, *3,* 066502.

[193] Versteega, D. J.; Bastiaansen, C. W. M.; Broer, D. J. *J. Appl. Phys.* 2002, *91*(71), 4191-4195.

[194] Chaudhari, P.; Lacey, J.; Doyle, J.; Galligan, E.; Lien, S.-C. A.; Callegari, A.; Hougham, G.; Lang, N. D.; Andry, P. S.; John, R.; Yang, K.-H.; Lu, M.; Cai, C.; Speidell, J.; Purushothaman, S.; Ritsko, J.; Samant, M.; Stöhr, J.; Nakagawa, Y.; Katoh, Y.; Saitoh, Y.; Sakai, K.; Satoh, H.; Odahara, S.; Nakano, H.; Nakagaki, J.; Shiota, Y. *Nature* 2001, *411*, 56-59.

[195] Stohr, J.; Samant, M. G.; Luning, J.; Callegari, A. C.; Chaudhari, P.; Doyle, J. P.; Lacey, J. A.; Lien, S. A.; Purushothaman, S.; Speidell, J. L. *Science* 2001, *292*, 2299-2302.

[196] Lee, H.-J.; Kim, Y.-H.; Lee, J.-J.; Park, H.-G.; Yang, S.; Kim, B.-Y.; Lee, Y. D.; Ju, B.-K.; Seo, D.-S. *Materials Chemistry and Physics* 2011, *126*, 628–631.

[197] Son, P. K.; Park, J. H.; Kim, J. C.; Yoon, T.-H.; Rho, S. J.; Shin, S. T.; Kim, J. S.; Lim, S. K.; Souk, J. H. *Jpn. J. Appl. Phys.* 2009, *48* 056503.

[198] Gwag, J. S.; Park, K.-H.; Kang, D. J.; Jhun, C. G.; Kim, H.; Cho, S. J.; Yoon, T.-H.; Kim, J. C. *Jpn. J. Appl. Phys.* 2003, *42,* L468.

[199] Seo, J.-H.; Lee, S. P.; Yoon, T.-H.; Kim, J. C. *Jpn. J. Appl. Phys.* 2007, *46* 6057.

[200] Wu, K. Y.; Chen, C.-H.; Yeh, C.-M.; Hwang, J.; Liu, P.-C.; Lee, C.-Y.; Chen, C.-W.; Wei, H. K.; Kou, C. S.; Lee, C.-D. *J. Appl. Phys.* 2005, *98*, 083518.

[201] Yaroshchuk, O. V.; Kravchuk, R. M.; Pogulay, S. S.; Tsiolko, V. V.; Kwok, H. S. *Applied Surface Science* 2011, *257,* 2443–2447.

[202] Oh, B.-Y.; Lee, K.-M.; Kim, B.-Y.; Kim, Y.-H.; Han, J.-W.; Han, J.-M.; Sang-Keuk Lee, S.-K.; Seo, D.-S. *J. Appl. Phys.* 2008, *104*, 064502.

[203] Gwag, J. S.; Jhun, C. G.; Kim, J. C.; Yoona, T.-H.; Lee, G.-D.; Cho, S. J. *J. Appl. Phys.* 2004, *96*(1), 257-260.

[204] Lee, K.-M.; Oh, B.-Y.; Kim, Y.-H.; Seo, D.-S. *J. Appl. Phys.* 2009, *105*, 014507.

[205] Wu, K. Y.; Chang, S. J.; Hwang, J.; Lee, C.-Y.; Tang, H.-C.; Chen, C.-W.; Liu, C. H.; Wei, H. K.; Kou, C. S.; Lee, C.-D. *J. Appl. Phys.* 2007, *101*, 023531.

[206] Matsumoto, T.; Kinoshita, Y.; Tanii, M.; Tatemichi, J.; Konishi, M.; Naito, M. CP1066, *Ion Implantation Technology,* edited by Seebauer, E. G.; Felch, S. B.; Jain, A.; Kondratenko, Y. V. 521-524.

[207] Kang, H.; Hwang, B. H.; Baik; H. K.; Kang, D.; Lee, J.-C. *Liquid Crystals* 2010, 37(2), 179–187.

[208] Lee, J.-W.; Moon, B.-M.; Lee,, K.-M.; Kim, Y.-H.; Park, H.-G.; Lim, J.-H.; Oh, B.-Y.; Kim, B.-Y.; Hwang, J.-Y.; Ok, C.-H.; Seo, D.-S.; Han, J.-M. *Liquid Crystals* 2010, *37*(3), 279–284.

[209] Wu, K.Y.; Hwang, J.; Lee, C.-Y.; Tang, H.-C.; Liu, Y.-L.; Liu, C.H.; Wei, H.K.; Kou, C.S. *Thin Solid Films* 2008, *517*, 905–908.

[210] Lee, J.-S.; Lee, C.-Y.; Kim, B.-S. *Surface & Coatings Technology* 2011, *206*, 986–989.

[211] Chien, H.W.; Wu, G.M.; Chiue, C.W.; Chen, C.N. *Surface & Coatings Technology* 2011, *206,* 797–800.

[212] Son, P. K.; Park, J. H.; Jo, B. K.; Lee, S. P.; Lee, J. H.; Kim, J. C.; Yoon, T.-H.; Lee, T. J.; Ree, M. *Thin Solid Films* 2009, *517*, 1803–1806.

[213] Park, H.-G.; Kim, Y.-H.; Kim, B.-Y.; Kim, D.-H.; Yoon, H.-I.; Seo, D.-S. *Thin Solid Films* 2011, *519,* 5654–5657.

[214] Suzuki, T.; Matsushima, J.; Sasaki, Y. *NEC Technical journal* 2006, *1*(3), 80-84.

[215] Kim, S. Y.; Park, H.-G.; Cho, M.-J.; Jeong, H.-C.; Kim, D.-H.; Seo, D.-S. *ECS Journal of Solid State Science and Technology* 2014, *3*(11) R212-R215.

[216] Yaroshchuk, O.; Kravchuk, R.; Dobrovolskyy, A.; Qiu, L.; Lavrentovich, O. *Liq.Cryst.* 2004, *31(*6), 859-869.
[217] Yaroshchuk, O.V.; Kravchuk, R.M.; Dobrovolskyy, A.M.; Liu, P.C.; Lee, C.D. *J. Soc. Inf. Display* 2005, *13*(4), 289-294.
[218] Yaroshchuk, O.; Kravchuk, R.; Parri, O. *SID'07 Digest* 2007, 694-697.
[219] Yaroshchuk, O.; Kravchuk, R.; Dobrovolskyy, A.; Pavlov, S. *Eurodisplay* 2002, 421-424.
[220] Choi, J.-H.; Ham, M.-H.; Oh, B.-Y.; Hwang, J.-Y.; Choi, S.-H.; Seo, D.-S.; Myoung, J.-M. *Liquid Crystals* 2006, *33*(8), 947–951.
[221] Chang, S. J.; Wu, K.Y.; Yang, Y.H.; Hwang, J.; Wu, H.-Y.; Pan, R.-P.; Lee, A.P.; Kou, C.S. *Thin Solid Films* 2007, *515*, 8000–8004.
[222] Wei, H.K.; Kou, C.S.; Wu, K.Y.; Hwang, J. *Diamond & Related Materials* 2008, *17*, 1639–1642.
[223] Baitukha, A.; Mori, S.; Suzuki, M. *Thin Solid Films* 2012, *523*, 37–40.
[224] Baitukha, A.; Mori, S.; Suzuki, M. *Thin Solid Films* 2014, *566*, 78–82.
[225] Lin , T.-C.; Yu, S.-C.; Chen, P.-S.; Chi, K.-Y.; Pan, H.-C.; Chao, C.-Y. *Current Applied Physics* 2009, *9*(3), 610–612.
[226] Kim, K.C.; Ahn,H.J.; Kim,J.B.; Hwang, B.K.; Baik,H.K. *Langmuir* 2005, 21(24), 11079-11084.
[227] Nemoto, F.; Nishiyama, I.; Takanishi, Y.; Yamamoto*, J. Soft Matter.* 2012, *8*, 11526-11530.
[228] Fukuhara, K.; Nagano, S.; Hara, M.; Seki, T. *Nature Comm.*2014, *5,* 1-8.
[229] Hwang, S.; Jin, H.-J.; Yoon, T.-H.; Kim, J. C. *Jpn. J. Appl. Phys.* 2010, *49*, 121702.
[230] Wang, W.; Sun, X.; Wu, W.; Peng, H.; Yu, Y. *Angewandte Chemie International Edition* 2012, 51(19), 4644–4647.
[231] Jeong, H. S.; Youn, S. C.; Kim, Y. H.; Jung, H.-T. *Phys. Chem. Chem. Phys.* 2013,*15*, 9493-9497.
[232] Russell, J. M.; Oh, S.; LaRue, I.; Zhou, O.; Samulski, E. T. *Thin Solid Films* 2006, *509*, 53 – 57.
[233] Tanner, J. R. *Novel Alignment Materials For Use in Liquid Crystal Displays,* PhD thesis, University of North Carolina at Chapel Hill, 2006.
[234] Hoogboom, J.; Garcia, P. M.; Otten, M. B.; Elemans, J. A.; Sly, J.; Lazarenko, S. V.; Rasing, T.; Rowan, A. E.; Nolte, R. J. *J. Am. Chem. Soc.* 2005, *127*(31), 11047-52.
[235] Hoogboom, J.; Elemans, J. A. A. W.; Rowan, A. E.; Rasing, T. H. M.; Nolte, R. J. M. *Phil. Trans. R. Soc. A* 2007, *365*, 1553–1576.
[236] Komitov, L. *Journal of the Society for Information Display* 2008, *16*(9), 919–925.
[237] Price, A. D.; Schwartz, D. K. *Langmuir* 2006, *22*(23), 9753-9.
[238] Malone, S. M.; Schwartz, D. K. *Langmuir* 2008, *24*(17), 9790-4.
[239] Suzuki, M.; Ferencz, A.; Iida, S.; Enkelmann, V.; Wegner, G. *Adv. Mater.* 1993, 5(5), 359–364.
[240] Price, A. D. Surface Anchoring of a Nematic Liquid Crystal at Solid and Fluid Interfaces, PhD thesis, University of Colorado, Boulder, CO, 2007.
[241] Zonga, Y.; Rqhec, J.; Knoll, W. *Thin Solid Films* 477 (2005) 203– 206.
[242] Vengatesan, M.R.; Lee, S.-H.; Son, J.-H.; Lim, J.-K.; Song, J. K. *Journal of Colloid and Interface Science* 2013, *407*, 310–317.
[243] Lu, R.; Xu, K.; Zhang, Y.; Gu, J.; Wu, H.; Lu, Z. *Physics Letters A* 1997, *231,* 449-452.

[244] Prompinit, P.; Achalkumar, A. S.; Bramble, J. P.; Bushby, R. J.; Walti, C.; Evans, S. D. *Applied Materials&Interfaces* 2010, 2(12), 3686–3692.
[245] Zhang, J.; Boamfa, M. I.; Rowan, A. E.; Rasing, T. *Adv. Mater.* 2010, *22*, 961–965.
[246] Fazio, V.S.U.; Komitov, L.; Lagerwall, S.T. *Thin Solid Films* 1998, *327–329*, 681–685.
[247] Kinsinger, M. I.; Buck, M. E.; Meli, M.-V.; Abbott, N. L.; Lynn, D. M. *Journal of Colloid and Interface Science* 2010, 341, 124–135.
[248] Zheng, W.; Chiang, C.-Y.; Underwood, I. *Thin Solid Films* 2013, *545*, 371–374.
[249] Komitov, L. *Journal of the Society for Information Display* 2003, *11*(3), 437-441.
[250] Yeung, F.S.Y.; Ho, Y.L.J.; Li,Y.W.; Kwok, H.S.; *J. Displ. Technol.* 2008, *4(1),* 24-27.

INDEX

D

E

F

M

N

O

P

Q

R

S

T

U

V

W

X

Y

Z